Recent Topics on Topology - From Classical to Modern Applications

Edited by Paul Bracken

Published in London, United Kingdom

Recent Topics on Topology - From Classical to Modern Applications
http://dx.doi.org/10.5772/intechopen.1003398
Edited by Paul Bracken

Contributors
Abderrahman El Boukili, Abhinav Prem, Alexis Chacón, Ghasem Mirhosseinkhani, Hicham Lekbich, Larry Li, Marcin Abram, Najim Mansour, Paul Bracken, Stephan Haas

First published in London, United Kingdom, 2024 by IntechOpen
IntechOpen is the global imprint of INTECHOPEN LIMITED, registered in England and Wales, registration number: 11086078, 167-169 Great Portland Street, London, W1W 5PF, United Kingdom

British Library Cataloguing-in-Publication Data
A catalogue record for this book is available from the British Library

Additional hard and PDF copies can be obtained from orders@intechopen.com

Recent Topics on Topology - From Classical to Modern Applications
Edited by Paul Bracken
p. cm.
Print ISBN 978-1-83769-838-7
Online ISBN 978-1-83769-837-0
eBook (PDF) ISBN 978-1-83769-839-4

Meet the editor

Paul Bracken is a professor in the Department of Mathematics, University of Texas Rio Grande Valley, USA. He received a BSc from the University of Toronto and a Ph.D. from the University of Waterloo, Canada. His research spans differential geometry, partial differential equations, and their applications to mathematical physics, with a focus on quantum mechanics and quantum field theory. He has published more than 210 research papers and books and has presented talks at numerous meetings and conferences. Dr. Bracken is a member of many societies including the American Mathematical Society (AMS), Canadian Mathematical Society (CMS), and American Physical Society (APS).

Contents

Preface

Topology plays a fundamental role in many areas of mathematics, such as differential geometry, analysis, and graph theory, in addition to being an independent area of study. It focuses on the properties of mathematical structures, such as point sets, that remain invariant under homeomorphisms. The chapters in this volume cover several topics that overlap with topology and its applications in physics. Topology is experiencing renewed interest in fields like condensed matter physics. Three chapters explore topology's applications in physics, while two chapters address modern usages of topology.

This book is written by an international group of invited authors. It is a pleasure to thank them for their hard work and significant contributions to this volume. I am also grateful to IntechOpen for the opportunity to edit this work and to Publishing Manager Ms. Tea Jelaca for her assistance throughout the process.

Paul Bracken
Professor,
Department of Mathematics,
University of Texas Rio Grande Valley,
Edinburg, TX, USA

Section 1

Applications of Topology in Modern Physics

Chapter 1

Hofstadter Butterflies in Topological Insulators

Larry Li, Marcin Abram, Abhinav Prem and Stephan Haas

Abstract

In this chapter, we investigate the energy spectra and the bulk and surface states in a two-dimensional system composed of a coupled stack of one-dimensional dimerized chains in the presence of an external magnetic field. Specifically, we analyze the Hofstadter butterfly patterns that emerge in a 2D stack of coupled 1D Su-Schrieffer-Heeger (SSH) chains subject to an external transverse magnetic field. Depending on the parameter regime, we find that the energy spectra of this hybrid topological system can exhibit topologically non-trivial bulk bands separated by energy gaps. Upon introducing boundaries into the system, we observe topologically protected in-gap surface states, which are protected either by a non-trivial Chern number or by inversion symmetry. We examine the resilience of these surface states against perturbations, confirming their expected stability against local symmetry-preserving perturbations.

Keywords: topological insulators, hybrid topology, surface states, topological protection, higher-order topology

1. Introduction

The classification and characterization of quantum phases of matter in terms of the topology of their many-body ground state(s) is one of the central concepts in modern condensed matter physics. The field of topological phases has witnessed remarkable synergy between theory and experiment, with the theoretical prediction and experimental discovery of electronic insulators [1–7] and superconductors [8–11] with topological band structures rapidly followed by the tenfold way classification of gapped phases of non-interacting fermions [12–14]. Given the spatial dimension, this classification specifies the distinct topological phases that are protected by the ten Altland-Zirnbauer symmetry classes [15]. Even within the restricted setting of non-interacting fermionic systems, it is by now well-understood that the non-trivial topology of electronic bands manifests in remarkable universal phenomena, including quantized topological invariants and robust gapless boundary modes which are stable to local symmetric perturbations and disorder [16–18].

More recently, the topological classification was extended to include systems with crystalline point-group symmetries, such as reflection or rotation. Generally speaking, gapped topological phases protected by such symmetries are referred to as topological crystalline phases and include a subset of higher-order topological insulators (HOTIs) and superconductors [19–28]. In contrast to conventional topological states in d spatial dimensions, which host protected gapless modes on their $d-1$ dimensional surfaces, an nth-order topological phase in d dimensions only host gapless modes on a $d-n$ dimensional surface and is gapped elsewhere. Such higher-order corner/hinge modes are correspondingly protected by higher-order bulk topological invariants, leading to a generalized notion of bulk-boundary correspondence [29]. Under the accepted framework for classifying quantum phases of matter, all of the above phases (topological insulators/superconductors, HOTIs, and Chern insulators) fall under the broad umbrella of *invertible* topological states, whose general classification includes the role of many-body correlations [30–35]. Invertible phases (a subset of which are so-called symmetry-protected topological (SPT) phases) are characterized by a trivial bulk (i.e., no fractionalized excitations above the ground state), a unique ground state on arbitrarily closed manifolds, and anomalous boundary modes [36].

Despite considerable progress in the classification of strongly correlated quantum matter, rich and interesting phenomena are still being uncovered in non-interacting electronic systems. In this chapter, we discuss the relatively new concept of "hybrid" topology [37–40] in a two-dimensional (2D) system. Specifically, we consider a 2D stack of dimerized one-dimensional (1D) chains that are coupled and subjected to an external magnetic field, resulting in a model that inherits the topological features of both the celebrated 1D Su-Schrieffer-Heeger (SSH) model [41] and the 2D Hofstadter model [42]. The former permits a topologically non-trivial phase, which is characterized by an integer bulk invariant and protected boundary modes; meanwhile, the latter describes charged particles hopping on a lattice in the presence of an external magnetic field and provides a lattice realization of the integer quantum Hall effect (IQHE), which exhibits a quantized Hall conductance and chiral edge modes. We show explicitly in this chapter that the combination of time-reversal symmetry breaking and dimerization leads to qualitatively different kinds of surface states protected either by a non-trivial Chern number or by inversion symmetry. However, rather than focusing on these bulk invariants, in this chapter, we will argue on general principles for the existence of edge states and demonstrate the presence of these states explicitly by exact diagonalization in the presence of a boundary. Our goal is to demonstrate the key non-trivial physics that emerges from these relatively simple models without invoking abstract mathematical concepts. In what follows, we first review, the essential background for the SSH and the Hofstadter models before considering the combination of the two, which will be the main focus of this chapter.

2. The Su-Schrieffer-Heeger model

We begin by considering the one-dimensional (1D) *Su-Schrieffer-Heeger (SSH)* model [41] and briefly review its topological phase diagram. The 1D SSH model is a bipartite chain with alternating hopping parameters, as illustrated in **Figure 1(a)**. Its Hamiltonian is given by

$$H_{SSH} = t_1 \sum_{m=1}^{N} \left(c^{\dagger}_{m,\mathrm{B}} c_{m,\mathrm{A}} + \mathrm{H.c.} \right) + t_2 \sum_{m=1}^{N-1} \left(c^{\dagger}_{m+1,\mathrm{A}} c_{m,\mathrm{B}} + \mathrm{H.c.} \right), \tag{1}$$

where N denotes the number of unit cells and A and B label the two sublattice sites within a unit cell m. Here, t_1 and t_2 are the intra- and inter-cell hopping parameters,

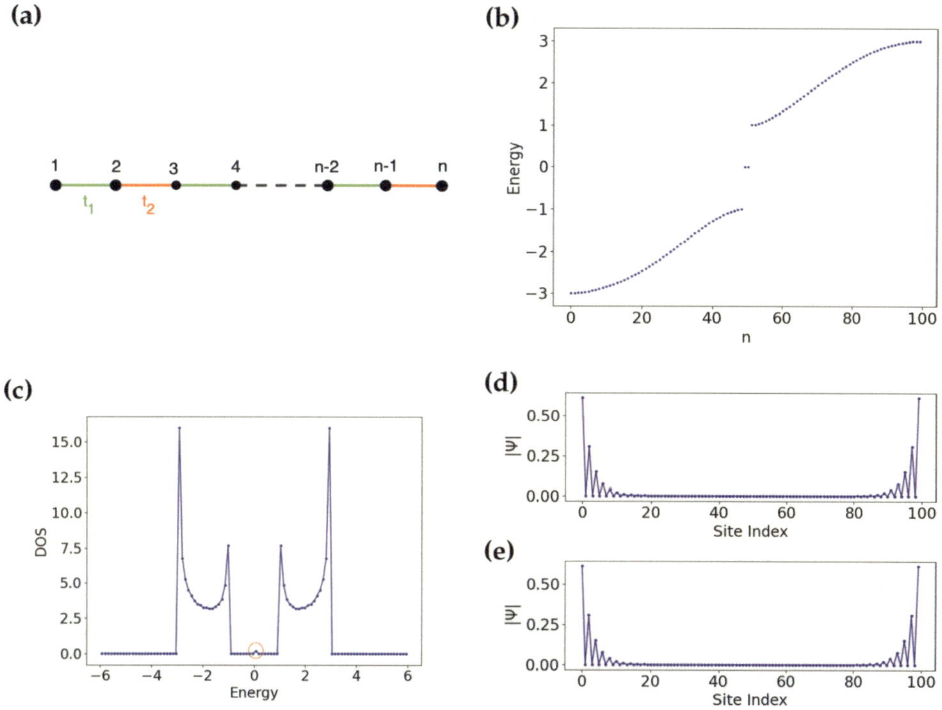

Figure 1.
Electronic structure of the Su-Schrieffer-Heeger (SSH) model. (a) Illustration of the real-space SSH model on a bipartite tight-binding chain. (b) Corresponding energy spectrum of the SSH chain for a 100-site chain with open boundary conditions (OBC). We observe two zero-energy edge states in the topological phase, whereas no zero-energy edge states are present in the trivial phase. (c) Corresponding density of states for a 2000-site chain featuring two bulk bands that are separated by a topological gap, with two degenerate surface states at E = 0. (d), (e)) Magnitude of the wave functions of the twofold degenerate zero-energy surface states, localized at the two chain edges.

respectively. In the presence of periodic boundary conditions (PBC), the Hamiltonian can be Fourier-transformed into momentum space. For each momentum k, we then have the single particle Hamiltonian [43].

$$H_{SSH}(k) = \begin{pmatrix} 0 & h^*(k) \\ h(k) & 0 \end{pmatrix}, \tag{2}$$

where $h(k) = h_x(k) + ih_y(k)$, with $h_x(k) = \text{Re}(t) + |t_2|\cos[ka + \arg(t_2)]$ and $h_y(k) = -\text{Im}(t) + |t_2|\sin[ka + \arg(t_2)]$. Here, a is the lattice spacing.

The bulk topological invariant for the SSH chain is given by the winding number $\mathcal{W}$, which can be evaluated via the momentum space integral:

$$\mathcal{W} = \frac{1}{2\pi i}\int_{-\pi}^{\pi} dk \frac{d}{dk} \ln[h(k)]. \tag{3}$$

In the following, we choose real-valued hopping parameters, that is, $t_1, t_2 \in \mathbb{R}$. For $t_1 > t_2$, we obtain $\mathcal{W} = 0$, that is, the system is in a topologically trivial phase, whereas for $t_1 < t_2$, the winding number evaluates to $\mathcal{W} = 1$, and the system is in the topologically non-trivial phase. The phase transition occurs at $t_1 = t_2$ at which point the bulk band gap closes and the winding number is ill-defined. This is consistent with the $\mathbb{Z}$ classification for 1D systems in the BDI symmetry class [12, 13].

Due to the bulk-boundary correspondence, the non-trivial topology of the SSH model is reflected in the number of pairs of zero-energy edge states $N_{\text{p-es}}$ in the case of open boundary conditions (OBC). These zero-energy edge states come in pairs due to the *chiral symmetry* of the SSH model, where the chiral symmetry operator is the Pauli matrix σ_z such that

$$\sigma_z H_{SSH}(k)\sigma_z = -H_{SSH}(k). \tag{4}$$

This immediately leads to the result that for any eigenvalue $E(k)$ of $H(k)$ with eigenvector $\Psi(k)$, there exists an orthogonal eigenvector $\sigma_z\Psi(k)$ with eigenvalue $-E(k)$. These two states are called chiral partners; in fact, the number of zero-energy edge state pairs in the SSH model is equal to the bulk winding number, that is, $N_{\text{p-es}} = \mathcal{W}$.

In **Figure 1(b)** we show the energy spectrum of the SSH chain in the topological ($t_1 = 0.75 < t_2 = 1.25$) phase for a system with 100 sites. The electronic structure in the trivial phase corresponds to a gapped particle-hole symmetric insulator with $N_{\text{p-es}} = 0$, whereas in the topologically non-trivial phase, we observe $N_{\text{p-es}} = 1$, that is, we find a pair of degenerate zero-energy electronic states as shown in **Figure 1(b)**. Moreover, the corresponding eigenstates for these zero-energy modes are exponentially localized at the edges of the chain, as shown explicitly in **Figure 1(d)** and **(e)**. In this case, these two zero-energy edge states are chiral partners of each other.

We now briefly review the two-dimensional (2D) extension of the SSH model (see Ref. [44] for more details). The 2D SSH model is defined on a 2D square-lattice with alternating hopping parameters in both the x- and the y-directions, as illustrated in **Figure 2(a)**. Consequently, there are four basis atoms: A, B, C, and D within a unit cell, and the intra- and inter-cell hopping parameters are again denoted by t_1 and t_2, respectively.

In contrast with conventional 2D topological insulators, the 2D SSH model has zero Berry curvature everywhere in the Brillouin zone due to the coexisting time-reversal and inversion symmetries. Hence, it is neither a Chern insulator nor is it a quantum spin Hall insulator; instead, the topological nature of this model can be characterized by the 2D Zak phase [44], which is given by

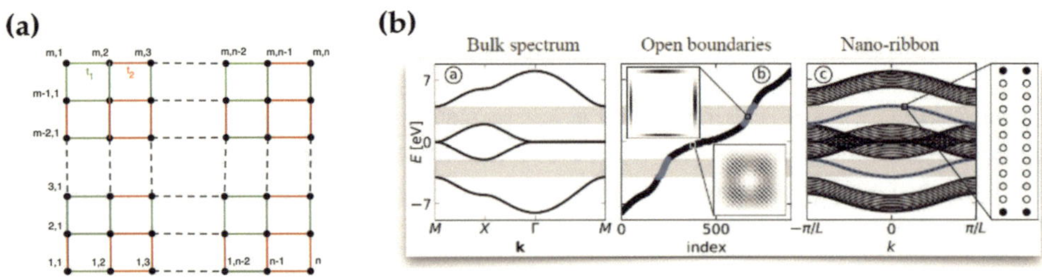

Figure 2.
Two-dimensional Su-Schrieffer-Heeger model. (a) Real-space illustration of the 2D SSH model on a square lattice. There are four basis atoms in each unit cell. The intra- and inter-cell hopping parameters are t_1 and t_2, respectively, in both the x- and y-directions. (b) Energy spectra for this system for different boundary conditions. For PBC in both directions, we only observe bulk states and the momenta k_x, k_y remain good quantum numbers. For OBC in both directions, we observe additional 0D and 1D topological surface states. For OBC in only one direction and PBC in the other direction, we encounter a nano-ribbon structure, with only 0D topological edge states.

$$\gamma = \int_{\mathrm{BZ}} d\vec{k}\, \mathrm{Tr}\left[\psi|i\nabla_{\vec{k}}|\psi\right] = \begin{cases} (0,0) & \text{if } t > t_2, \\ (\pi,\pi) & \text{if } t < t_2. \end{cases} \tag{5}$$

The topologically non-trivial phase of the 2D SSH model nevertheless displays in-gap edge and corner localized modes. In particular, 0D corner states and 1D edge states exist along the open boundaries of a square lattice, as well as on the 0D end points and 1D edges of a nano-ribbon structure. The corresponding energy spectra for these two cases are shown in **Figure 2(b)**.

3. The Hofstadter model

In this section, we turn to the problem of an electron hopping on a 2D square lattice in the presence of a transverse magnetic field, which provides a non-interacting lattice model for a Chern insulator. The presence of a lattice introduces competition between the lattice spacing a and the magnetic length ℓ_0 (set by the magnetic field), with the incommensuration between these two length scales leading to the paradigmatic Hofstadter butterfly [42].

3.1 Electron in a magnetic field

First, we briefly review the canonical example of a charged particle moving in a 2D plane subject to a uniform transverse magnetic field $\vec{B} = B\hat{z}$, with dynamics governed by the Hamiltonian:

$$\hat{H} = \frac{1}{2M}\left(-i\hbar\nabla + \frac{e}{c}\mathbf{A}\right)^2, \tag{6}$$

where M is the mass of the electron and $\mathbf{A}$ is the vector potential. The net flux through the system is $\Phi = BL_1L_2 = N_\phi\phi_0$, where L_1 and L_2 are the dimensions along $\hat{x}$ and $\hat{y}$ respectively, $\phi_0 = hc/e$ is the flux quantum, and N_ϕ denotes the number of flux quanta through the plane. Defining the magnetic length $\ell_0 = \sqrt{\frac{\hbar c}{eB}} = \sqrt{\frac{\phi_0}{2\pi B}}$ and working in units where $\hbar = c = e = 1$, we have

$$\phi_0 = 2\pi; N_\phi = \frac{L_1L_2}{2\pi\ell_0^2}; \ell_0 = \sqrt{\frac{1}{B}}. \tag{7}$$

It is straightforward to diagonalize this Hamiltonian (see, e.g., Ref. [45]), whose energy levels are the degenerate Landau levels $E_n = \omega_c\left(n + \frac{1}{2}\right)$, where $n \in \mathbb{Z}_{\geq 0}$ is the Landau level (LL) index and $\omega_c = 1/\left(M\ell_0^2\right)$ is the cyclotron frequency.

The macroscopic degeneracy of the LLs can be understood in terms of the magnetic translation algebra, as first noted by Zak [46]. Due to the vector potential, the naïve translation operator $e^{i\mathbf{p}\cdot\mathbf{R}}$ does not commute with $\hat{H}$. We can instead introduce the generator of infinitesimal magnetic translations $\mathbf{k} = \mathbf{p} - \mathbf{A}$, with the magnetic translation operator given by $\hat{T}(\mathbf{R}) = e^{i\mathbf{k}\cdot\mathbf{R}}$. These operators form a projective representation of the group of magnetic translations

$$\hat{T}(\mathbf{R}_1)\hat{T}(\mathbf{R}_2) = \exp\left(\frac{i}{2\ell_0^2}(\mathbf{R}_1 \times \mathbf{R}_2)\cdot\hat{z}\right)\hat{T}(\mathbf{R}_1 + \mathbf{R}_2) \tag{8}$$

where $\hat{z}$ is normal to the plane. The Hamiltonian, in terms of the conjugate momentum operator, $\pi = \mathbf{p} + \mathbf{A}$ then takes the simple form: $\hat{H} = \pi^2/(2M)$. Working in the symmetric gauge, $A = \frac{1}{2}\mathbf{B} \times \mathbf{r}$, it is clear that the magnetic translation operators commute with the Hamiltonian $[k_i, \hat{H}] = 0$ but not among themselves since $[k_i, k_j] = -\frac{i}{\ell_0^2}\epsilon_{ij}\,(i,j = 1,2)$. Thus, we can use either k_1 or k_2 to label the degenerate LL states. To explicitly see this degeneracy, define

$$\hat{T}_x = \hat{T}\left(\frac{L_1}{N_\phi}\hat{x}\right);\ \hat{T}_y = \hat{T}\left(\frac{L_2}{N_\phi}\hat{y}\right), \tag{9}$$

which satisfies the magnetic algebra

$$\hat{T}_x\hat{T}_y = e^{i\frac{2\pi}{N_\phi}}\hat{T}_y\hat{T}_x. \tag{10}$$

(Note that this encodes the fact that the full symmetry group in the presence of a magnetic field is a central extension of translations $\mathbb{Z}^2$ by U(1) charge conservation.) Now suppose ψ is a simultaneous eigenstate of $\hat{H}$ and $\hat{T}_x$ such that $\hat{H}\psi = E\psi, \hat{T}_x\psi = t\psi$. Then, the state $\psi_m = \hat{T}_y^m\psi$ is an orthogonal eigenstate, with

$$\hat{H}\psi_m = E\psi_m;\ \hat{T}_x\psi_m = te^{i\frac{2\pi m}{N_\phi}}\psi_m, \tag{11}$$

with the degeneracy of each LL lower bounded by the number of flux quanta N_ϕ.

3.2 Square lattice in a magnetic field

We now consider the same problem as before but now regularized on a square lattice. Setting the lattice spacing $a = 1$, the Hofstadter model is defined as

$$\hat{H}_H = -t\sum_{<i,j>}\left[c_i^\dagger c_j \exp\left(i\int_{\mathbf{r}_i}^{\mathbf{r}_j}\mathbf{A}\cdot d\mathbf{l}\right) + \text{h.c}\right], \tag{12}$$

where $c^\dagger, c$ are fermionic creation/annihilation operators that satisfy canonical anti-commutation relations, i,j label lattice sites ($i = (m,n)$), and A encodes the uniform transverse magnetic field B. The lattice translation operators are

$$\hat{t}_x = \sum_{m,n} c_{m+1,n}^\dagger c_{m,n}\exp\left(i\int_m^{m+1}\mathbf{A}\cdot d\mathbf{x}\right), \hat{t}_y = \sum_{m,n} c_{m,n+1}^\dagger c_{m,n}\exp\left(i\int_n^{n+1}\mathbf{A}\cdot d\mathbf{y}\right), \tag{13}$$

which satisfies the algebra $\hat{t}_y\hat{t}_x = e^{i\phi}\hat{t}_x\hat{t}_y$, where $\phi = Ba^2$ is the uniform flux per plaquette. In terms of these operators, $\hat{H} = -t(\hat{t}_x + \hat{t}_y + \text{h.c})$, which does not commute with the naïve translation operators. The incommensurability between a and ℓ_0 is apparent in the translation algebra, since $\frac{\phi}{\phi_0} = \frac{a^2}{2\pi\ell^2}$.

In analogy with the continuum, we now look for operators that commute with the Hamiltonian. Working in the Landau gauge $\mathbf{A} = B(0, x, 0)$, the magnetic translation operators are

$$\hat{T}_x = \sum_{m,n} c^{\dagger}_{m+1,n} c_{m,n} e^{in\phi}; \hat{T}_y = \sum_{m,n} c^{\dagger}_{m,n+1} c_{m,n}. \tag{14}$$

One can verify that these commute with $\hat{H}$ and satisfy the magnetic translation algebra $\hat{T}^q_x \hat{T}_y = e^{i\phi q} \hat{T}_y \hat{T}^q_x$. Note that for rational flux per plaquette $\phi = \frac{p}{q}\phi_0$, where p and q are co-prime integers, the two operators $\hat{T}^q_x$ and $\hat{T}_y$ commute. As these are still translation operators–$\hat{T}^q_x$ translates a single particle state by q in the x-direction and $\hat{T}_y$ by 1 in the y-direction–we can label their eigenstates with momentum quantum numbers

$$\hat{T}^q_x |k_x, k_y\rangle = e^{iqk_x} |k_x, k_y\rangle, \hat{T}_y |k_x, k_y\rangle = e^{ik_y} |k_x, k_y\rangle. \tag{15}$$

Thus, in the presence of a magnetic field and for rational $\phi/\phi_0 = p/q$, the unit cell is enlarged. Here, the magnetic unit cell is $q \times 1$ lattice plaquettes; equivalently, the wave vectors belong to the magnetic Brillouin zone (mBZ) $k_x \in \left[-\frac{\pi}{q}, \frac{\pi}{q}\right]; k_y \in [-\pi, \pi]$. Finally, to impose PBC on this system, we require $L_1 = rq$ where $r \in \mathcal{Z}$. The total flux through the system, $\phi_{\text{total}} = \phi L_1 L_2 = \phi_0 (rpL_1L_2)$ is thus an integer multiple of the flux quanta.

In analogy with the continuum case, we expect degenerate energy levels. Consider an eigenstate $|k_x, k_y\rangle$ of $\hat{H}$. Since $\left[\hat{H}, \hat{T}_x\right] = 0$, $\hat{T}_x |k_x, k_y\rangle$ is a degenerate eigenstate; however,

$$\hat{T}_y \hat{T}_x |k_x, k_y\rangle = e^{-i\phi} \hat{T}_x \hat{T}_y |k_x, k_y\rangle = e^{i(k_y - \phi)} \hat{T}_x |k_x, k_y\rangle. \tag{16}$$

Since $\phi = 2\pi p/q$ (with p, q co-prime) and as $T_x |k_x, k_y\rangle = \ |k_x, k_y - \phi\rangle$ has the same energy, the spectrum is at least q fold degenerate.

Moving away from the square lattice, we now introduce spatial anisotropy into the system, with the Hamiltonian (in the Landau gauge) given by

$$\hat{H}_H = \sum_{m,n} \left[-t_a c^{\dagger}_{m+1,n} c_{m,n} - t_b c^{\dagger}_{m,n+1} c_{m,n} e^{2\pi i \Phi m} + \text{h.c}\right] \tag{17}$$

where $\alpha = p/q$ (with p, q co-prime integers) is the finite fraction of the flux quanta $\phi_0 = 2\pi$ through each plaquette. Since we have an enlarged $q \times 1$ magnetic unit cell, we expect q energy bands. To see this, we move to Fourier-space via

$$c_{m,n} = \frac{1}{(2\pi)^2} \int_{-\pi}^{\pi} \mathrm{dk_x} \int_{-\pi}^{\pi} \mathrm{dk_y} e^{i(k_x m + k_y n)}, \tag{18}$$

and use the magnetic translation symmetry of the system, which results in a mBZ that is q-times smaller in the x-direction:

$$\begin{aligned} \hat{H}_H = \frac{1}{(2\pi)^2} \int_{-\frac{\pi}{q}}^{\frac{\pi}{q}} \mathrm{dk_x} \int_{-\pi}^{\pi} \mathrm{dk_y} \sum_{m=0}^{q-1} \Big[-2t_a \cos(k_x + 2\pi\Phi m) c^{\dagger}_{m,k_y} c_{m,k_y} \\ -t_b \left(e^{ik_y} c^{\dagger}_{m-1,k_y} c_{m,k_y} + e^{-ik_y} c^{\dagger}_{m+1,k_y} c_{m,k_y}\right)\Big]. \end{aligned} \tag{19}$$

To simple notations, we denote $c_{m,k_y} \equiv c_{k_x+2\pi\Phi m,k_y}$. This is a 1D tight-binding model with an enlarged unit cell and a $q \times q$ Bloch Hamiltonian with q energy levels. The energy levels are obtained by considering the action of $\hat{H}_H$ on the single particle state $\psi = \sum_{m=0}^{q-1} \psi_m c^\dagger_{m,k_y}$ which leads to the Harper equation [47].

$$v_m \psi_{m,k_y} - t_b \left(e^{ik_y} \psi_{m+1,k_y} + e^{-ik_y} \psi_{m-1,k_y} \right) = E\psi_{m,k_y}; \quad v_m = -2t_a \cos(k_x + 2\pi\Phi m), \tag{20}$$

subject to $\psi_{m+q,k_y} = \psi_{m,k_y}$. In the Landau gauge, all eigenstates have trivial y-dependence, and we can write $\psi_{m,k_y} = e^{-ik_y m}\psi_m$, leading to

$$v_m \psi_m - t_b (\psi_{m+1} + \psi_{m-1}) = E\psi_m, \tag{21}$$

subject to the constraint that $\psi_{m+q} = e^{ik_y q}\psi_m$. From this, we recover the q-fold degeneracy of the spectrum

$$E(k_x, k_y) = E\left(k_x, k_y + \frac{2\pi m}{q}\right); m \in \mathbb{Z}, m \in [0, q-1]. \tag{22}$$

In the fully anisotropic case $t_a = 1, t_b = 0$, the magnetic field does not play a role as it only enters through the hopping in the y-direction (in the Landau gauge). The spectrum, in this case, is simply given by $-2t\cos(k_x)$ and can be understood as the dispersion in the full BZ folded onto the magnetic BZ. For the fully isotropic case $t_a = t_b = 1$, we show the spectrum $E(k_x, k_y = 0)$ at a fixed k_y in **Figure 3** for two cases: $p = 2, q = 3$ and $p = 3, q = 4$. In general, when q is even, the spectrum is symmetric about $E = 0$ while for q odd, the $E = 0$ level lies at the center of the middle band. Importantly, the system has $q - 1$ gaps when q is odd. Another interesting feature [48] is that for q even, there exist zero-energy solutions when

$$2t_a^q \cos(qk_x) + 2t_b^q \cos\left(qk_y\right) = (-1)^{\frac{q}{2}}(t_a^q + t_b^q), \tag{23}$$

such the momenta at which the energy vanishes are given by $k_x = 0, k_y = 2\pi m/q$ (when $q \in 4\mathbb{Z}$) and $k_x = \pi/q, k_y = (2m+1)\pi/q$ (when $q \in 4\mathbb{Z} + 2$), for

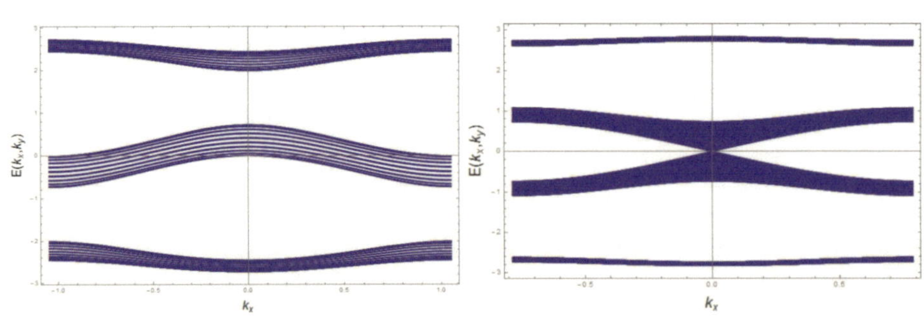

Figure 3.
Energy spectrum at fixed k_y momenta $(k_y = 0)$ for the isotropic Hofstadter model $(t_a = t_b = 1)$ with rational flux per plaquette $\phi = p/q\phi_0$: (a) $p = 2, q = 3$ and (b) $p = 3, q = 4$. The former has a gapped spectrum while the latter exhibits degenerate Dirac nodes at $E = 0$.

$m \in \mathbb{Z}, m \in [0, q-1]$. The system thus exhibits q Dirac nodes which form when the bands touch around $E = 0$. In the vicinity of these nodes, the dispersion takes the relativistic form $E \propto \pm q\sqrt{t_a^q k_x^2 + t_b^q k_y^2}$, which can be seen explicitly in **Figure 3(b)**, where $q = 4$.

Finally, we note that the full solution to Harper's equation, originally found by Hofstadter [42], results in the famed Hofstadter butterfly, which is found by plotting the energy spectrum E as a function of the flux through each plaquette α. We reproduce this spectrum for both PBC and OBC in **Figure 4**, where the gapped regions observed in the former setting are seen to host surface states upon introducing boundaries into the system.

3.3 Edge states on the square lattice

The Hall conductance for the Hofstadter model can be computed by evaluating the TKNN invariant [49] within the gapped regions of the butterfly; as expected for a Chern insulator, the Hall conductance $\sigma_{xy} = ne^2/h$ $(n \in \mathbb{Z})$ is quantized, with different values of n corresponding to different topological phases that are separated by gap closing transitions. Moreover, as expected by the bulk-boundary correspondence, a non-zero n implies the existence of gapless states localized at physical boundaries of the system. This is easily understood as follows: consider a non-trivial insulator $(n \neq 0)$ which shares a boundary with a trivial insulator $(n = 0)$. Since the hall conductance is a topological invariant, it can only change if the bulk gap closes, and so as long as the edge of the system respects U(1) charge conservation (required to define the Hall conductance), the only way for σ_{xy} to change across the boundary is if there exists a state localized on the edge that crosses the Fermi level.

The original argument for the existence of edge states was given by Laughlin [50, 51]. Consider placing a non-trivial Chern insulator on a cylindrical geometry, through which Φ flux quanta are threaded (in the longitudinal direction). Now, suppose a voltage V_x is applied transverse to k_y. Since the adiabatic insertion of a flux is the same as adiabatically switching on a vector potential, the flux shifts the momentum $k_y \to k_y + 2\pi\frac{\Phi}{L}$, such that $I_y = \Delta E / \Delta\Phi$ (via the Byers-Yang formula),

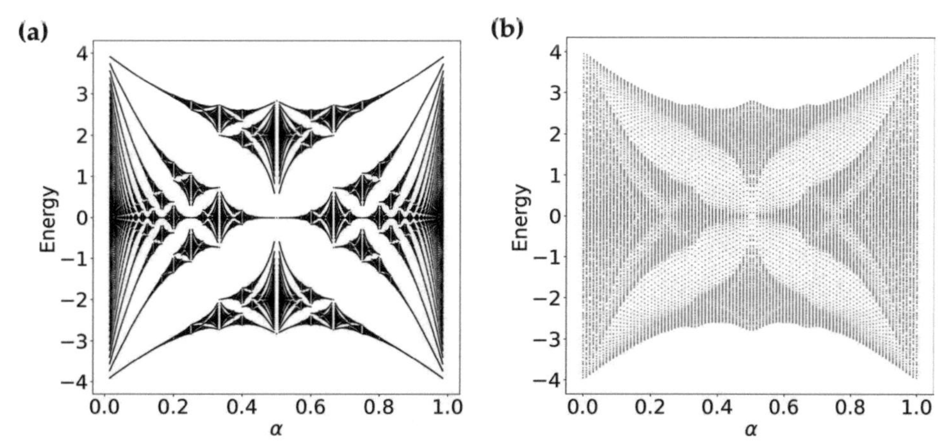

Figure 4.
Hofstadter Butterfly spectrum on a 20 × 20 tight-binding square lattice: (a) using PBC and rational momenta, corresponding to magnetic flux per plaquette in integer multiples of the magnetic flux quantum; (b) using OBC in real space, leading to surface states in the otherwise "forbidden" regions.

where ΔE is the change in energy during the process of flux insertion. Suppose that the Fermi level crosses some bands and that precisely one flux quanta is adiabatically threaded through the cylinder. The momentum of all states then changes precisely $\frac{2\pi}{L_y}$, which (due to PBC) is the spacing between momentum states. Thus, for all of the bands that the Fermi level crosses, there will be one state occupied above the Fermi level close to one edge and one state empty below the Fermi level close to the opposite edge. If the Fermi level crosses n bands, this implies that $\Delta E = nV_x$, since the two edges have a potential difference of V_x. Thus, $\sigma_{xy} = \frac{I_y}{V_x} = n$. Since this argument depends only on gauge invariance, the result is robust even in the presence of disorder.

For concreteness, consider again the Hofstadter model. We consider a cylindrical geometry, where we impose PBC in the y-direction (with length L_y) but OBC in the x-direction. The two edges are fixed at $x = 0$ and $x = L_x$, such that the system has $L_x - 1$ sites along the x-direction. We then assume that Φ flux quanta are threaded through this cylinder. In a slight abuse of notation, we now call the fraction of flux per plaquette ϕ ($\phi = \frac{Ba^2}{\phi_0}$ where B is the magnetic field). The Hamiltonian, again in the Landau gauge $\left(A_y = B\hat{x}\right)$ is

$$\hat{H}_H = -\sum_{m,n}\left[t_a c^\dagger_{m+1,n}c_{m,n} + t_b e^{2\pi i \frac{\Phi}{L_y}} c^\dagger_{m,n+1}c_{m,n}e^{2\pi i\phi m} + \text{h.c}\right]. \tag{24}$$

Due to the finite geometry, we no longer have translation invariance in the x-direction, and hence k_x is no longer a good quantum number. Fourier transforming in the y-direction, $c_{m,n} = \frac{1}{\sqrt{L_y}}\sum_k e^{ik_y n}c_m(k)$, with $k = \frac{2\pi m}{L_y}$ $\left(m = 1, 2, \dots, L_y\right)$, we find

$$\hat{H}_H = -\sum_{m,k}\left[t_a c^\dagger_{m+1,k}c_{m,k} + t_b e^{2\pi i \frac{\Phi}{L_y}} e^{i(2\pi\phi m - k)} c^\dagger_{m,k}c_{m,k}\text{h.c}\right]. \tag{25}$$

The action of this Hamiltonian on a single particle state, $|\Psi(k,\Phi)\rangle = \sum_m \psi_m(k,\Phi)$ results in

$$-t_a\left(\psi_{m+1} + \psi_{m-1}\right) - 2t_b\cos\left(k - 2\pi\frac{\Phi}{L_y} - 2\pi\phi m\right)\psi_m = E\psi_m. \tag{26}$$

This equation can equivalently be written in matrix form and studied within the transfer matrix formalism developed by Hatsugai [52]. While analytic results, such as the existence of edge modes, can be obtained in the case where L_x is commensurate with q (i.e., $L_x = rq$ with $r \in \mathbb{Z}$), in general, this problem can only be studied numerically. Here, we exactly diagonalize the preceding 1D Hamiltonian and present the numerical results in **Figure 5** for the isotropic case with $p = 2, q = 5$. In each case, the edge states connect the different bulk bands and are localized on the $x = 0$ and $x = L_x$ edges. Note that while the bulk states are indifferent to commensurability and the Hall conductance is the same in either case, the nature of the edge states is different [52].

4. Hofstadter butterflies in coupled SSH chains

In this section, we investigate a hybrid model in which features of the 1D SSH model coexist with those of the Hofstadter model. We note that analogous models

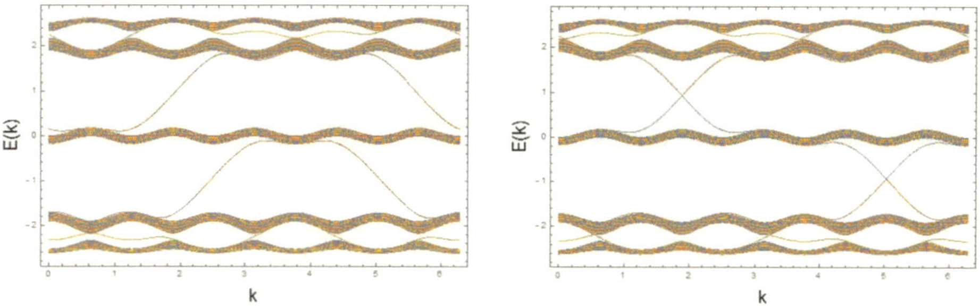

Figure 5.
Edge states in the Hofstadter model on a cylindrical geometry. Parameters here are $t_a = t_b = 1$, $p = 2$, and $q = 5$. On the left, $L_x = 120$ (commensurate with q), while on the right $L_x = 121$ (incommensurate with q).

have previously been considered in Refs. [37, 38, 53, 54]. Specifically, we consider a stack of 1D SSH chains that are coupled together to form a 2D lattice that is further subjected to a uniform transverse magnetic field. The underlying Hamiltonian for this system is given by

$$H = -t_0\sum_{i,j}\left(e^{i\phi_i}c^\dagger_{i,j}c_{i,j+1} + h.c.\right) - t_1\sum_{i,j}\left(c^\dagger_{2i-1,j}c_{2i,j} + h.c.\right) - t_2\sum_{i,j}\left(c^\dagger_{2i,j}c_{2i+1,j} + h.c.\right), \tag{27}$$

where the site positions (i,j) are defined on a 2D lattice of m coupled chains with n sites each; that is, we are considering m-leg ladders. The horizontal intra-chain coupling is staggered in accordance with the SSH Hamiltonian, whereas the vertical inter-chain coupling is modulated by the flux to incorporate the external magnetic field in accordance with the Hofstadter Hamiltonian (see **Figure 6** for an illustration). Since our focus will be on topological surface states, we will primarily consider systems with open boundary conditions (OBC).

To get an intuition for this model, consider the case where no magnetic field is switched on. Then, we do not expect a non-trivial Chern number in this system, which we have also verified explicitly. Moreover, if we place the system on a cylinder, with PBC along the y-direction and OBC along the x-direction, we expect topological surface states that are inherited from the underlying SSH chains when $t_2 > t_1$ and no such states when $t_2 < t_2$. The coupling between the chains then delocalizes these 0D states along the edge, leading to topologically protected modes that disperse along the 1D surface. In fact, it is straightforward to analyze the model on this geometry and show that the edge states have a dispersion $\sim t_0 \cos(k_y)$ and that they are protected by an effectively 1D winding invariant $\mathcal{W}(k_y)$ (evaluated at fixed k_y) that is non-trivial when $t_2 > t_1$. Once we turn on the external magnetic field, the system can now additionally host a non-trivial Chern number (strictly speaking, this will only be quantized in the 2D limit, i.e., as we take the number of legs m to be large), which can be calculated via the TKNN invariant [49]. Thus, we expect the distinct phases of this system to be characterized by two independent topological invariants: the Chern number and an effective 1D winding invariant. To evaluate the latter, the system is placed on a cylinder geometry with PBC along y, resulting in an effectively 1D Hamiltonian parameterized by k_y. Depending on the magnetic flux, this effective 1D system may preserve inversion symmetry $(\mathcal{I}H(k)\mathcal{I}^{-1} = H(-k))$ for various values of

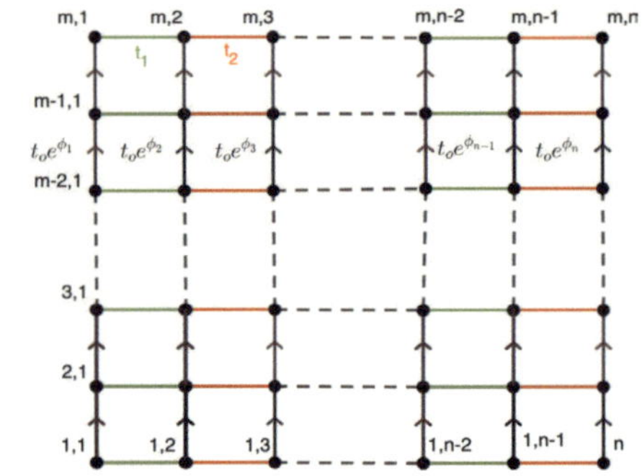

Figure 6.
Illustration of an m-leg SSH ladder of chains with n sites. The horizontal intra-chain couplings, t_1 and t_2, have alternating strength. The vertical inter-chain couplings, $t_0 \exp(i2\pi\alpha x)$, where x is the coordinate in the horizontal direction, are modulated by the external magnetic field, parameterized by $\alpha = Ba^2e/h$.

k_y [53], and can hence lead to an integer invariant and corresponding inversion symmetry-protected edge modes.

Let us first consider the minimal system in which we may observe the interplay between dimerization and the magnetic field, that is, the 2-leg ladder, since a single chain does not offer any plaquettes that can accommodate non-trivial magnetic flux. In the absence of a magnetic field ($\phi = 0$), the phase diagram of this hybrid model on a two-leg ladder consists of a gapped, topologically trivial phase (when $t_1 > t_2$, $t_1 + t_2 < t_0$ and $t_0 < t_1 - t_2$), a gapless critical region (when $|t_1 + t_2| > t_0 > |t_1 - t_2|$), and a gapped topologically non-trivial dimerized phase (when $t_2 > t_1$, $t_2 + t_1 < t_0$ and $t_0 < t_2 - t_1$). The latter phase inherits its non-trivial topology from the underlying SSH chain and is characterized by localized topological surface states whose eigenenergies lie within the bulk gap. Since this system is still quasi-1D, we do not expect a sharply quantized Chern number.

We now numerically investigate the Hofstadter butterfly spectrum in the two-leg SSH ladder, shown in **Figure 7**, for system dimensions $m = 2$ and $n = 50$ with OBC. In the upper panel, we compare the energy spectrum of the critical, isotropic system ($t_0 = t_1 = t_2 = 1$) in **Figure 7(a)** with that of the topologically non-trivial dimerized phase ($t_0 = t_1 = 1, t_2 = 2$) in **Figure 7(b)**. In both cases, the overall bandwidth of the spectrum is given by $W = t_0 + 2(t_1 + t_2)$, and the latter clearly displays bulk energy bands separated by an energy gap. In addition, in the topologically non-trivial dimerized phase, we observe a pair of twofold degenerate states that lie within the bulk gap and which are spatially localized at the two open ends of the ladder system.

This phenomenology persists into the regime where the inter-chain coupling dominates, as shown in **Figure 7(c)** and **(d)**. In the trivially dimerized phase ($t_0 = 1, t_1 = 0.2, t_2 = 0.1$) there are two bulk bands but no surface states within the energy gap separating them (see **Figure 7(c)**). In contrast, in the topologically non-trivial dimerized phase ($t_0 = 1, t_1 = 0.1, t_2 = 0.2$), additional gaps open up within the bulk bands, with twofold degenerate non-dispersive topological surface states appearing in the center of these gaps at energies $\pm t_0$.

The situation is different in the regime of dominant intra-chain interactions. As seen in **Figure 7(e)** and **(f)**, the spectrum is quasi-continuous in the trivially dimerized phase, whereas in the non-trivial phase, a central gap emerges, again with a pair of twofold degenerate surface states at energies $\pm t_0$. Also, note that for the 2-leg

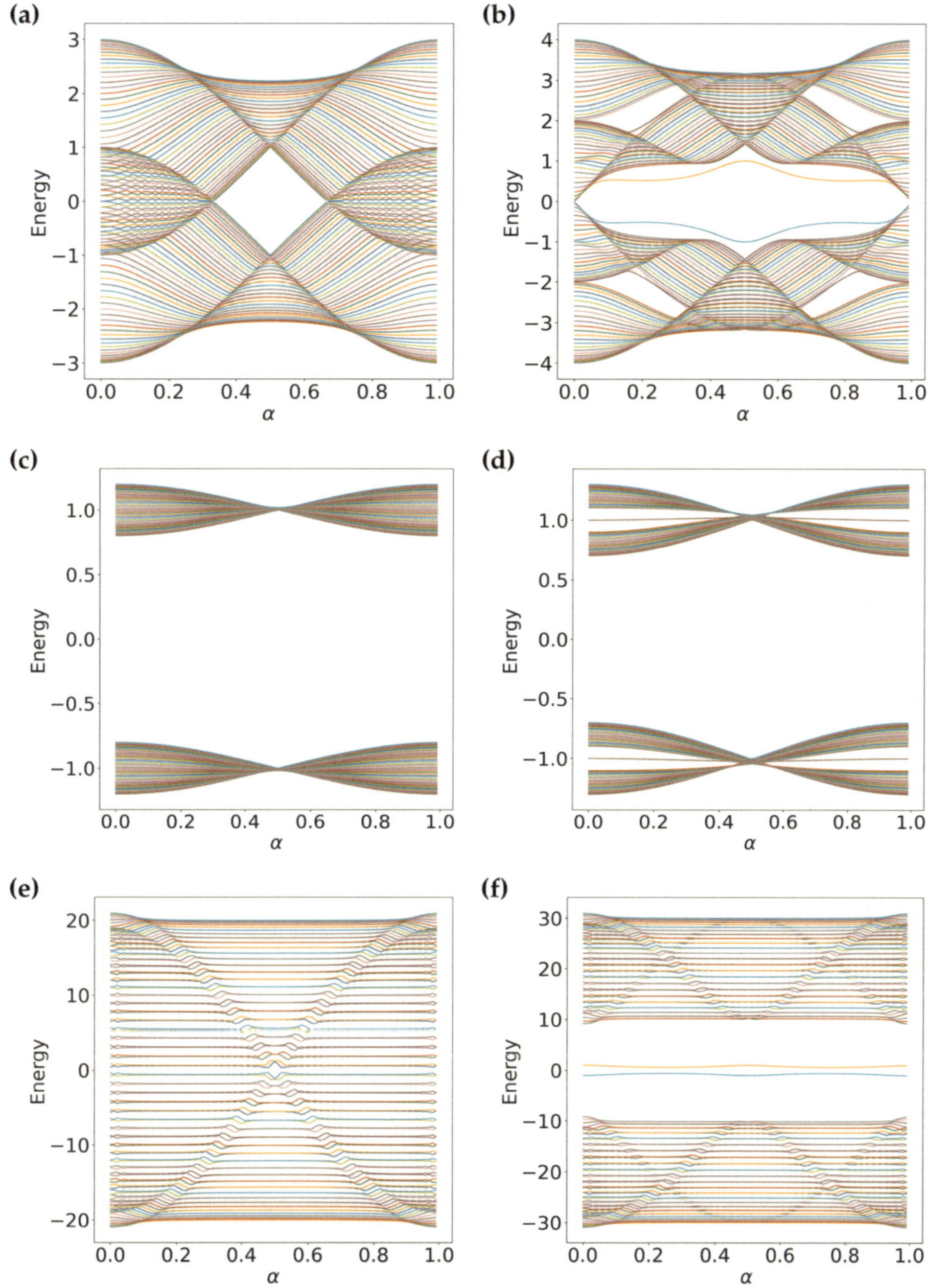

Figure 7.
Hofstadter butterfly spectra of the two-leg SSH ladder. (a) Critical point – isotropic limit, $t_0 = t_1 = t_2 = 1$; (b) Topologically non-trivial dimerized phase with four twofold degenerate surface states, $t_2 = 2$, $t_1 = t_0 = 1$; (c) Trivially dimerized phase without surface states, dominant inter-chain coupling, $t_1 = 0.2$, $t_2 = 0.1$, $t_0 = 1$; (d) Topologically non-trivial dimerized phase, dominant inter-chain coupling, $t_1 = 0.1$, $t_2 = 0.2$, $t_0 = 1$; (e) Trivially dimerized phase without surface states, dominant intra-chain couplings, $t_1 = 20$, $t_2 = 10$, $t_0 = 1$; (f) Topologically non-trivial dimerized phase, dominant intra-chain couplings, $t_1 = 10$, $t_2 = 20$, $t_0 = 1$.

SSH ladder, all of the surface states are zero-dimensional and are located at the corners of the system.

Next, we turn to the 4-leg ladder system and examine which of the above features carry over as we begin approaching the 2D limit. The corresponding Hofstadter spectra are shown in **Figure 8**. In **Figure 8(a)** and **(b)**, we compare the trivially dimerized and topologically non-trivial dimerized phases for the case when all hopping parameters are of the same order of magnitude. In contrast to the two-leg ladder, we now observe multiple gapped regions in the spectrum, some of which correspond to magnetic flux-induced states with non-trivial Chern numbers. While this can be checked by direct calculation of the TKNN invariant, the non-trivial topology is also reflected in the presence of topological surface states within bulk gaps due to the bulk-boundary correspondence (recall that we are using OBC here). Here, we observe that in the topologically non-trivial dimerized phase, there are surface states not only in the central gapped region (i.e., around $\alpha = 0.5$) but also within other gaps. Within the central bulk gap, we find a pair of twofold degenerate surface states, which are localized at the two horizontal ends of the ladder. More generally, if we consider, say $\alpha = 0.25$, we observe in-gap surface states both at 1/4-filling and also at 1/2-filling. The former are chiral edge states that correspond to a non-zero Chern number $C = 1$ in that gap, while the latter are protected by inversion symmetry of an effective 1D Hamiltonian [53] but not by a Chern number, as that is trivial within this region of the butterfly.

For the case where the energy scales are well-separated, that is, either dominant inter-chain couplings (**Figure 8(c)** and **(d)**) or dominant intra-chain couplings (**Figure 8(e)** and **(f)**), the resulting Hofstadter spectra are a straightforward generalization of what is observed in the two-leg ladder. Namely, for $t_0 \gg t_1, t_2$, there are now four instead of two bulk bands. In the topologically non-trivial phase, gaps open up in the center of these bands, and a twofold degenerate surface state emerges in each of these. In the opposite limit, $t_0 \ll t_1, t_2$, we observe two bulk bands, same as for the two-leg ladder, with four twofold degenerate surface states within the central bulk gap.

We now consider the 2D case, with the numerical results for a 20×20 system with OBC shown in **Figure 9**. In the trivially dimerized phase, shown in **Figure 9(a)**, we recognize the slightly altered well-known pattern of the Hofstadter spectrum in an isotropic 2D tight-binding lattice. Due to the OBC, the bulk-gap regions of the periodic system are populated by chiral surface states, which result from the non-trivial Chern number in these regions, which will be further discussed below. In the topologically non-trivial dimerized phase, shown in **Figure 9(b)**, we observe additional surface states that are not protected by a Chern number (which vanishes in this region of the butterfly) but are instead protected by the inversion symmetry of an effective 1D Hamiltonian (see discussion in Ref. [53]).

Moving on to the cases of dominant inter-chain (**Figure 9(c)** and **(d)**) and intra-chain (**Figure 9(e)** and **(f)**) couplings, we observe that the spectra in the trivially dimerized regimes look quite similar in the 2D limit, although the overall energy scales are either set by t_0 or by t_1 and t_2. In both cases, one observes an emerging quasi-continuum. However, what is different for these two cases are the energy levels of the topological surface states in the non-trivial regimes: for dominant inter-chain couplings (**Figure 9(c)**(d)), they emerge within each of these 20 mini-bands, whereas for dominant intra-chain couplings ((**Figure 9(f)**), they form a central $E \approx 0$ surface state band, separated from the bulk bands by a gap of order $t_2 - t_1$.

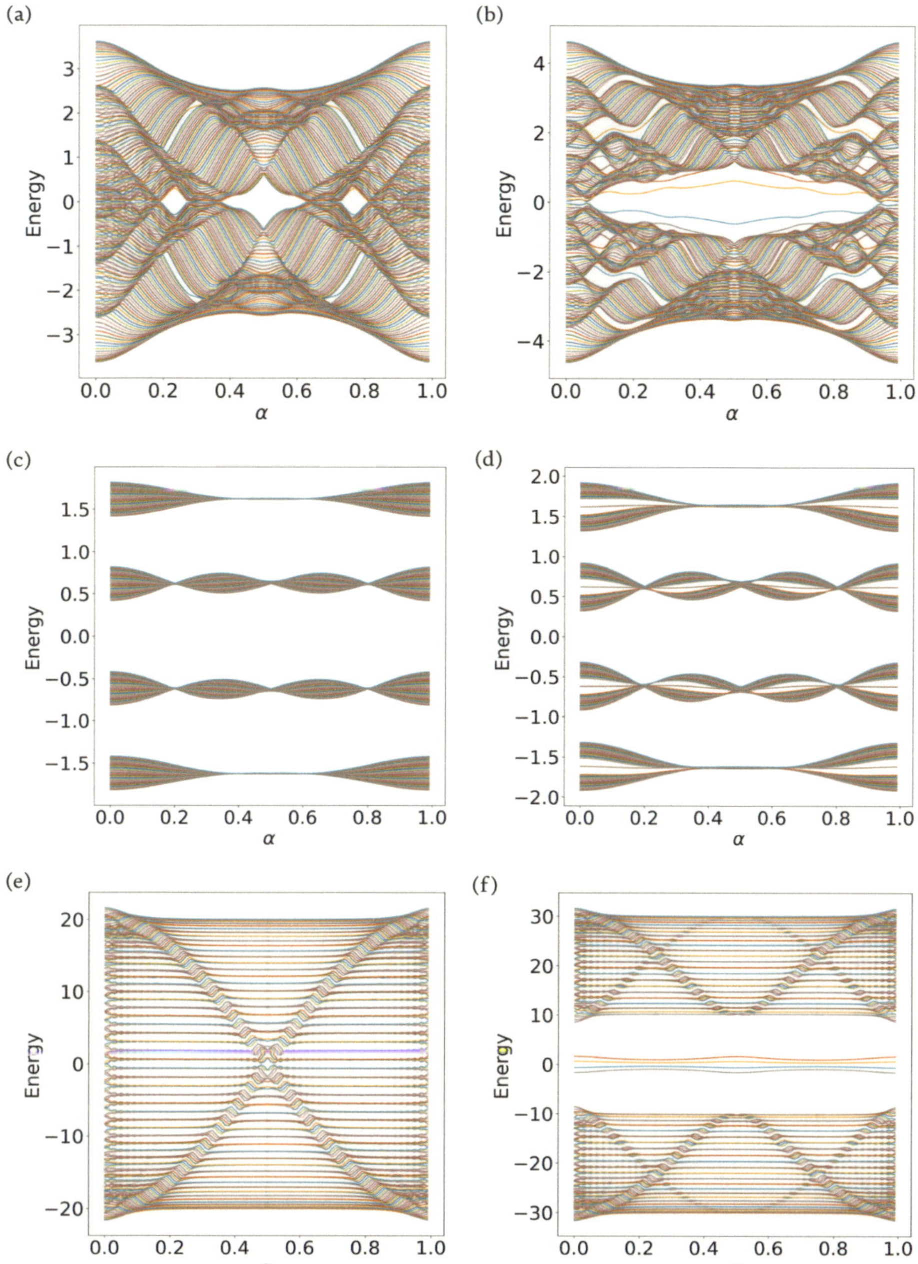

Figure 8.
Hofstadter butterfly spectra of the four-leg SSH ladder. (a) Trivially dimerized phase without surface states, $t_0 = 1, t_2 = 1, t_1 = 2$; (b) Topologically non-trivial dimerized phase with eight twofold degenerate surface states, $t_2 = 2$, $t_1 = t_0 = 1$; (c) Trivial dimerized phase without surface states, dominant inter-chain coupling, $t_1 = 0.2$, $t_2 = 0.1$, $t_0 = 1$; (d) Topologically non-trivial dimerized phase, dominant inter-chain coupling, $t_1 = 0.1$, $t_2 = 0.2$, $t_0 = 1$; (e) Trivially dimerized phase without surface states, dominant intra-chain couplings, $t_1 = 20$, $t_2 = 10$, $t_0 = 1$; (f) Topologically non-trivial dimerized phase, dominant intra-chain couplings, $t_1 = 10$, $t_2 = 20$, $t_0 = 1$.

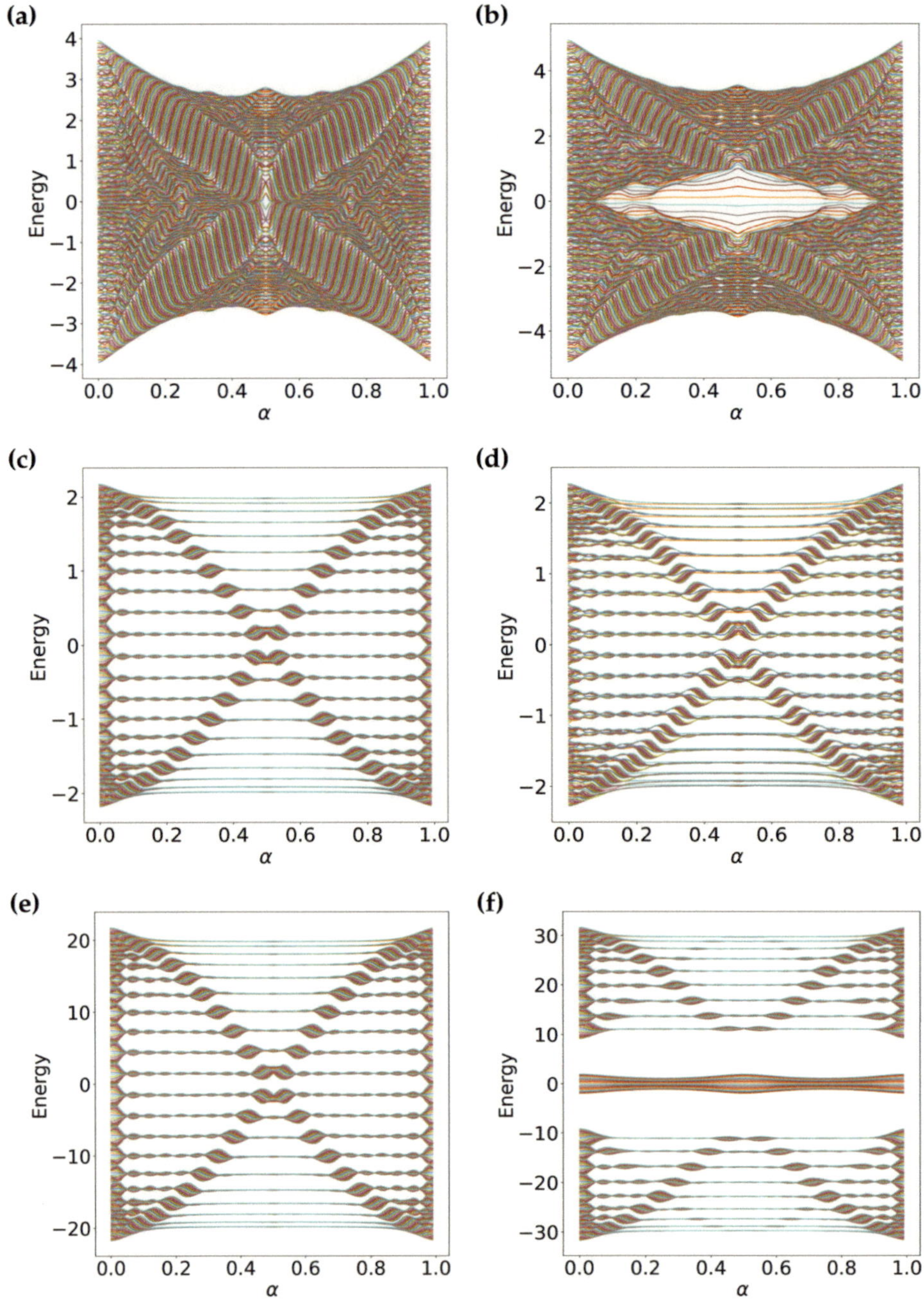

Figure 9.
Hofstadter butterfly spectra of the 20-leg SSH ladder. (a) Trivially dimerized phase, $t_0 = t_2 = 1$, $t_1 = 2$; (b) Topologically non-trivial dimerized phase with ten 4-fold degenerate surface states in the central bulk gao, $t_2 = 2$, $t_1 = t_0 = 1$; (c) Trivially dimerized phase without surface states, dominant inter-chain coupling, $t_1 = 0.2$, $t_2 = 0.1$, $t_0 = 1$; (d) Topologically non-trivial dimerized phase, dominant inter-chain coupling, $t_1 = 0.1$, $t_2 = 0.2$, $t_0 = 1$; (e) Trivially dimerized phase without surface states, dominant intra-chain couplings, $t_1 = 20$, $t_2 = 10$, $t_0 = 1$; (f) Topologically non-trivial dimerized phase, dominant intra-chain couplings, $t_1 = 10$, $t_2 = 20$, $t_0 = 1$.

We now move on to a discussion of the nature of the various surface states that can be observed in this hybrid topological system. Here, we focus on the 20 × 20 system, but the following observations are applied more generally. In **Figure 10**, we first focus on the surface states with energies within the central bulk gap, that is, around $\alpha = 0.5$ and $|E| < 1$. We have highlighted 4 of these in **Figure 10(a)**. These disperse only slightly as a function of the applied magnetic field, and, as seen in **Figure 10(b)-(e)**, they are localized at the two ends of the coupled ladder systems perpendicular to the direction of the magnetic vector potential. Also, note that these surface states come in degenerate pairs, with their number determined by a 1D topological invariant of an inversion symmetric effective 1D Hamiltonian [53].

In contrast, consider the other type of surface states, which result from a non-trivial Chern number and are shown in **Figure 11**. Again, we have highlighted 4 of these states, this time in the gapped region within the upper left lobe of the Hofstadter butterfly spectrum, **Figure 11(a)**. As seen in **Figure 11(b)-(e)**, these surface states are chiral (since they correspond to a non-trivial Chern number) and are qualitatively different from the ones shown in **Figure 10**. They are localized at all four edges of the system and also, they do not display the spatial modulations observed in **Figure 10(b)-(e)**.

Finally, we examine how these different topological surface states respond to the introduction of a local perturbation. Specifically, we consider the effects of an onsite potential $U = 1$, introduced at the lower left corner of the system. This potential obviously breaks the spatial symmetries of the system, with time-reversal symmetry further broken by the magnetic field. As seen in **Figure 12(a)**, this type of impurity breaks the twofold degeneracy of the inversion symmetry-protected surface states. As illustrated in **Figure 12(b)-(e)**, the corresponding wave functions are now localized only on one side of the system, thus adapting to the local onsite repulsion.

In contrast, the onsite repulsive potential has no discernible effect on the chiral edge states, which stem from the non-trivial Chern number, as shown in **Figure 13**. While this is clearly expected, given the nature of this hybrid topological system, it also points to potential applications that allow us to address and tune some of the prominent topological surface states, while leaving others unaltered.

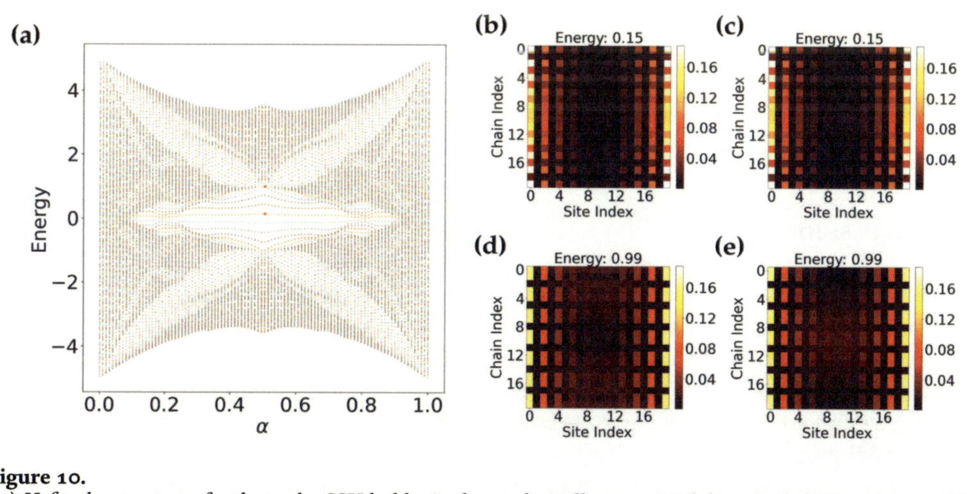

Figure 10.
(a) Hofstadter spectrum for the 20-leg SSH ladder in the topologically non-trivial dimerized phase with ten 4-fold degenerate surface states protected by inversion symmetry, $t_2 = 2$, $t_1 = t_0 = 1$; (b) and (c) two of the four degenerate topological surface states at energy E = 0.15; (d) and (e) two of the four degenerate topological surface states at energy E = 0.99.

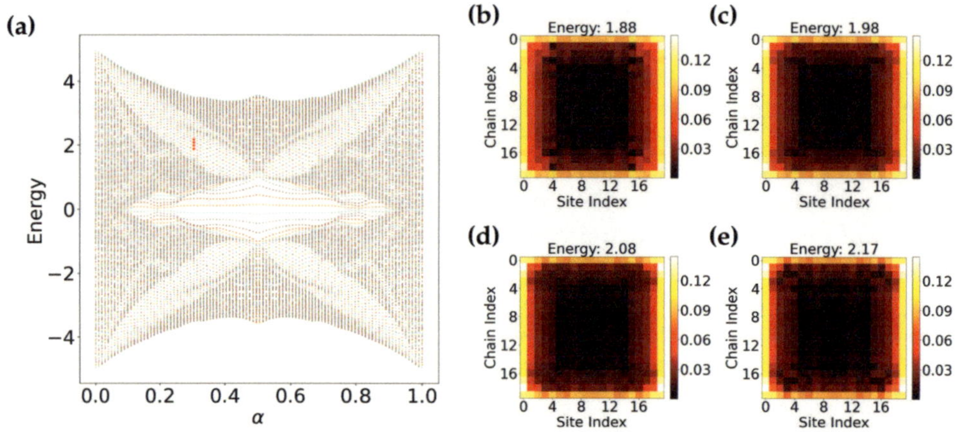

Figure 11.
(a) Hofstadter spectrum of the 20-leg SSH ladder in the topologically non-trivial dimerized phase with non-degenerate surface states due to non-trivial Chern number, $t_2 = 2$, $t_1 = t_0 = 1$; *(b) topological surface states at energy E = 1.88; (c) E = 1.98, (d) E = 2.08, and (e) E = 2.17.*

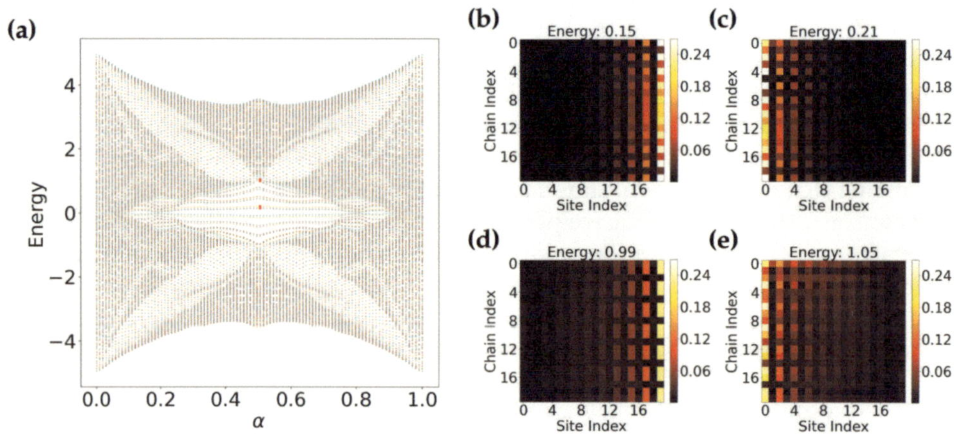

Figure 12.
Effect of an onsite impurity (strength U = 1) on the inversion symmetry-protected surface states in the topologically non-trivial dimerized phase. The impurity is located in the lower left corner of the 20-leg SSH ladder. Parameters are chosen identically to the previous figure.

5. Conclusions

In this chapter, we have explored the intricate energy spectra and the behavior of bulk and surface states in topological insulators characterized by coexisting topological invariants, with a particular focus on the Hofstadter butterfly patterns that emerge. Through our analysis of coupled stacks of 1D Su-Schrieffer-Heeger (SSH) chains, we have shown that the interplay of topologically non-trivial dimerization and an external magnetic field can lead to distinct kinds of surface states.

In these hybrid topological systems, we have observed well-defined bulk energy bands that are well-separated by energy gaps. As is expected from the conventional Hofstadter model, one can calculate the Hall conductance for this system and find distinct phases that are characterized by distinct integer-valued topological invariants,

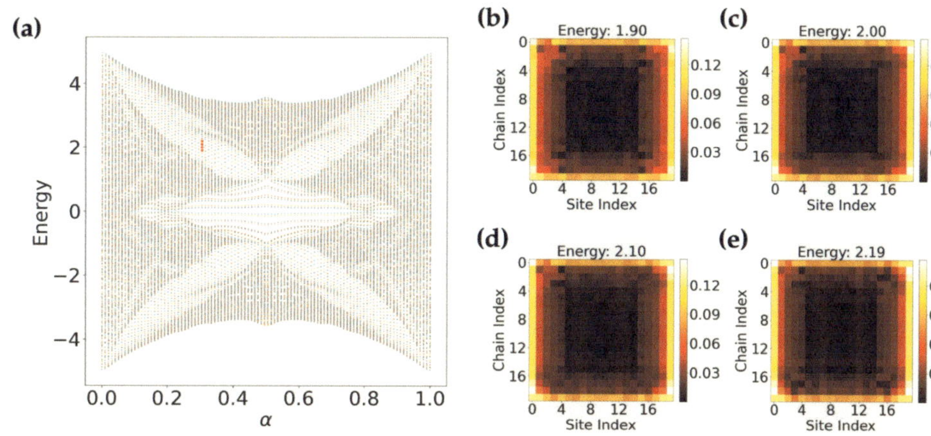

Figure 13.
No effect of an onsite impurity (strength U = 1) on the surface states due to the non-trivial Chern number in the topologically non-trivial dimerized phase. The impurity is located in the lower-left corner of the 20-leg SSH ladder. Parameters are chosen identical to the previous figure. 20 × 20 *square lattice,* $H[0][0] = 1$ *lattice* $t_0 = 1, t_1 = 1, t_2 = 2$ *at index* 30.

that is, Chern numbers. Upon introducing boundaries into the system, these non-trivial Chern numbers result in chiral edge states, which appear within the bulk gaps and are localized on the boundaries of the system. Additionally, however, the topologically non-trivial dimerization introduced by the underlying SSH model can also lead to topologically non-trivial surface states which are protected by the inversion symmetry of an effective 1D model and a corresponding 1D topological invariant (as discussed in Ref. [53]). This protection ensures the resilience of these states against various perturbations, highlighting the crucial role of symmetry in maintaining their stability.

Our examination of the SSH and Hofstadter models has revealed that the nature of the symmetry-protected surface states differs significantly between these models. The SSH model, which preserves inversion symmetry, and the Hofstadter model, which breaks time-reversal symmetry and hence can have a non-trivial Chern number, each giving rise to distinct types of surface states. These states are not only localized differently but also exhibit different responses to perturbations, underscoring the diverse ways in which topological protection can manifest.

More generally, we can also imagine introducing dimerization in the vertical direction, which would lead to the coexistence of 1D edge states and 0D corner states within the same system: as such, this would constitute an instance of a hybrid higher-order topological insulator in which 0D corner states, non-chiral 1D edge states, and chiral 1D edge states coexist [37, 38]. This illustrates the rich tapestry of topologically protected boundary states that can emerge upon subjecting spatial symmetry-protected topological states to an external magnetic field.

In conclusion, our study highlights the fascinating physics that arises from combining strong topology (as encoded in the Chern number) and crystalline symmetries, which can lead to weak topological invariants. By investigating relatively simple, yet paradigmatic models, we have illustrated how hybrid topology can lead to the emergence of robust surface states, further enriching our understanding of topological phases of matter. We hope that the insights gained from this work lead to further exploration of hybrid topological systems and their potential applications.

Author details

Larry Li[1], Marcin Abram[1], Abhinav Prem[2] and Stephan Haas[1*]

1 Department of Physics and Astronomy, University of Southern California, Los Angeles, CA, USA

2 School of Natural Sciences, Institute for Advanced Study, Princeton, New Jersey, USA

*Address all correspondence to: shaas@usc.edu

References

[1] Kane CL, Mele EJ. Z_2 topological order and the quantum spin hall effect. Physical Review Letters. 2005;**95**:146802

[2] Andrei Bernevig B, Zhang S-C. Quantum spin hall effect. Physical Review Letters. 2006;**96**:106802

[3] Liang F, Kane CL, Mele EJ. Topological insulators in three dimensions. Physical Review Letters. 2007;**98**:106803

[4] Moore JE, Balents L. Topological invariants of time-reversal-invariant band structures. Physical Review B. 2007;**75**:121306

[5] Roy R. Topological phases and the quantum spin hall effect in three dimensions. Physical Review B. 2009;**79**: 195322

[6] König M, Wiedmann S, Brüne C, Roth A, Buhmann H, Molenkamp LW, et al. Quantum spin hall insulator state in HgTe quantum Wells. Science. 2007; **318**(5851):766

[7] Hsieh D, Qian D, Wray L, Xia YQ, Hor YS, Cava RJ, et al. A topological dirac insulator in a quantum spin hall phase. Nature. 2008;**452**(7190):970

[8] Read N, Green D. Paired states of fermions in two dimensions with breaking of parity and time-reversal symmetries and the fractional quantum hall effect. Physical Review B. 2000;**61**: 10267-10297

[9] Ivanov DA. Non-abelian statistics of half-quantum vortices in p-wave superconductors. Physical Review Letters. 2001;**86**:268-271

[10] Stone M, Roy R. Edge modes, edge currents, and gauge invariance in $p_x + ip_y$ superfluids and superconductors. Physical Review B. 2004;**69**:184511

[11] Zhang P, Yaji K, Hashimoto T, Ota Y, Kondo T, Okazaki K, et al. Observation of topological superconductivity on the surface of an iron-based superconductor. Science. 2018;**360**(6385):182-186

[12] Ryu S, Schnyder AP, Furusaki A, Ludwig AW. Topological insulators and superconductors: Tenfold way and dimensional hierarchy. New Journal of Physics. 2010;**12**(6):065010

[13] Kitaev A. Periodic table for topological insulators and superconductors. AIP Conference Proceedings. 2009;**1134**(1):22-30

[14] Chiu C-K, Teo JCY, Schnyder AP, Ryu S. Classification of topological quantum matter with symmetries. Reviews of Modern Physics. 2016;**88**: 035005

[15] Altland A, Zirnbauer MR. Nonstandard symmetry classes in mesoscopic normal-superconducting hybrid structures. Physical Review B. 1997;**55**:1142-1161

[16] Hasan MZ, Kane CL. Colloquium: Topological insulators. Reviews of Modern Physics. 2010;**82**:3045-3067

[17] Qi X-L, Zhang S-C. Topological insulators and superconductors. Reviews of Modern Physics. 2011;**83**:1057-1110

[18] Zahid Hasan M, Moore JE. Three-dimensional topological insulators. Annual Review of Condensed Matter Physics. 2011;**2**:55-78

[19] Liang F. Topological crystalline insulators. Physical Review Letters. 2011; **106**:106802

[20] Hsieh TH, Lin H, Liu J, Duan W, Bansil A, Liang F. Topological crystalline insulators in the SnTe material class. Nature Communications. 2012;**3**:982

[21] Okada Y, Serbyn M, Lin H, Walkup D, Zhou W, Dhital C, et al. Observation of dirac node formation and mass acquisition in a topological crystalline insulator. Science. 2013; **341**(6153):1496-1499

[22] Sessi P, Di Sante D, Szczerbakow A, Glott F, Wilfert S, Schmidt H, et al. Robust spin-polarized midgap states at step edges of topological crystalline insulators. Science. 2016;**354**(6317):1269-1273

[23] Ma J, Yi C, Lv B, Wang ZJ, Nie S, Wang L, et al. Experimental evidence of hourglass fermion in the candidate nonsymmorphic topological insulator khgsb. Science Advances. 2017;**3**: e1602415

[24] Schindler F, Cook AM, Vergniory MG, Wang Z, Stuart SP, Parkin B, et al. Higher-order topological insulators. Science Advances. 2018;**4**: eaat0346

[25] Wladimir A, Benalcazar B, Bernevig A, Hughes TL. Quantized electric multipole insulators. Science. 2017;**357**(6346):61-66

[26] Langbehn J, Peng Y, Trifunovic L, von Oppen F, Brouwer PW. Reflection-symmetric second-order topological insulators and superconductors. Physical Review Letters. 2017;**119**:246401

[27] Song Z, Fang Z, Fang C. $(d-2)$-dimensional edge states of rotation symmetry protected topological states. Physical Review Letters. 2017;**119**:246402

[28] Khalaf E, Po HC, Vishwanath A, Watanabe H. Symmetry indicators and anomalous surface states of topological crystalline insulators. Physical Review X. 2018;**8**:031070

[29] Khalaf E, Benalcazar WA, Hughes TL, Queiroz R. Boundary-obstructed topological phases. Physical Review Research. 2021;**3**:013239

[30] Chen X, Zheng-Cheng G, Liu Z-X, Wen X-G. Symmetry protected topological orders and the group cohomology of their symmetry group. Physical Review B. 2013;**87**:155114

[31] Zheng-Cheng G, Wen X-G. Symmetry-protected topological orders for interacting fermions: Fermionic topological nonlinear σ models and a special group supercohomology theory. Physical Review B. 2014;**90**:115141

[32] Song H, Huang S-J, Liang F, Hermele M. Topological phases protected by point group symmetry. Physical Review X. 2017;**7**:011020

[33] Huang S-J, Song H, Huang Y-P, Hermele M. Building crystalline topological phases from lower-dimensional states. Physical Review B. 2017;**96**:205106

[34] Thorngren R, Else DV. Gauging spatial symmetries and the classification of topological crystalline phases. Physical Review X. 2018;**8**:011040

[35] Else DV, Huang S-J, Prem A, Gromov A. Quantum many-body topology of quasicrystals. Physical Review X. 2021;**11**:041051

[36] Senthil T. Symmetry-protected topological phases of quantum matter. Annual Review of Condensed Matter Physics. 2015;**6**(1):299-324

[37] Otaki Y, Fukui T. Higher-order topological insulators in a magnetic field. Physical Review B. 2019;**100**:245108

[38] Zuo Z-W, Benalcazar WA, Liu Y, Liu C-X. Topological phases of the dimerized Hofstadter butterfly. Journal of Physics D Applied Physics. 2021; **54**(41):414004

[39] Hossain MS, Schindler F, Islam R, Muhammad Z, Jiang Y-X, Cheng Z-J, et al. A hybrid topological quantum state in an elemental solid. Nature. 2024; **628**(8008):527-533

[40] Huang S-J, Park K, Hsu Y-T. Hybrid-order topological superconductivity in a topological metal 1T'-$MoTe_2$. npj Quantum Materials. 2024;**9**:21

[41] Su WP, Schrieffer JR, Heeger AJ. Solitons in polyacetylene. Physical Review Letters. 1979;**42**:1698-1701

[42] Hofstadter DR. Energy levels and wave functions of bloch electrons in rational and irrational magnetic fields. Physical Review B. 1976;**14**: 2239-2249

[43] Asbóth JK, Oroszlány L, Pályi A. A short course on topological insulators. Lecture Notes in Physics. 2016;**919**:166

[44] Liu F, Wakabayashi K. Novel topological phase with a zero berry curvature. Physical Review Letters. 2017; **118**:076803

[45] MacDonald AH. Introduction to the physics of the quantum hall regime. arXiv e-prints. 1994:cond–mat/9410047

[46] Zak J. Magnetic translation group. Physics Review. 1964;**134**:A1602-A1606

[47] Harper PG. Single band motion of conduction electrons in a uniform magnetic field. Proceedings of the Physical Society A. 1955;**68**(10): 874-878

[48] Wen XG, Zee A. Winding number, family index theorem, and electron hopping in a magnetic field. Nuclear Physics B. 1989;**316**(3):641-662

[49] Thouless DJ, Kohmoto M, Nightingale MP, den Nijs M. Quantized hall conductance in a two-dimensional periodic potential. Physical Review Letters. 1982;**49**:405-408

[50] Laughlin RB. Quantized hall conductivity in two dimensions. Physical Review B. 1981;**23**:5632-5633

[51] Halperin BI. Quantized hall conductance, current-carrying edge states, and the existence of extended states in a two-dimensional disordered potential. Physical Review B. 1982;**25**: 2185-2190

[52] Hatsugai Y. Edge states in the integer quantum hall effect and the riemann surface of the bloch function. Physical Review B. 1993;**48**:11851-11862

[53] Lau A, Ortix C, van den Brink J. Topological edge states with zero hall conductivity in a dimerized hofstadter model. Physical Review Letters. 2015; **115**:216805

[54] He A-L, Zhang X, Liu Y. Topological states in a dimerized square-octagon lattice with staggered magnetic fluxes. Physical Review B. 2022;**106**:125147

Chapter 2

Topological String Theory: A Source for Conformal Field Theories

Abderrahman El Boukili, Hicham Lekbich and Najim Mansour

Abstract

In this paper, we first recall the approach to extract Liouville field theories from the action of worldsheets superstring in the standard case and also in the presence of gauge fields. In order to extend this formalism, we treat the case of Toda field equations based on a special ansatz of our superfields. Then, we study the integrability of the system by the Lax formulation based on the structure of the associated Lie superalgebra.

Keywords: topological string theory, gauged superstring, Liouville field theories, Toda field equation, integrability

1. Introduction

It is commonly known that bosonic string theory has several problems. In particular, it is tachyonic, and it does not have fermionic states describing matter [1, 2]. It cannot be used to unify particle physics and gravity. To overcome such a problem, we should add fermions on the worldsheet ψ^{μ} [3, 4]. To make this realization, we can introduce fermionic fields into the Ramond–Neveu–Schwarz string theory [5, 6]. However, the action of this RNS model is obtained by extending the Polyakov action to a 2D worldsheet action with a local supersymmetry [7–9]. Consequently, the action obtained has in addition to Weyl symmetries a local superconformal symmetries [10, 11].

In 1967, Leonard Susskind is the first one to introduce the worldsheet surface as a generalization of the worldline describing the motion of a point particle in special and general relativity, and it is seen in topological string theory as a two-dimensional manifold that describes the motion of the string in spacetime [12, 13].

However, when we are only interested in the topology of the worldsheet, the resulting theory is called the topological string theory where only the variables of the worldsheet enter in the calculation of the associated partition function [14, 15]. Consequently, the topological string theory corresponds in the case of the conformal theory coupled to gravity to a nonlinear two-dimensional integrable model [16, 17].

In the present paper, we would like to introduce a new ansatz on the 2D superstring worldsheet action in order to give the super-Liouville and super Toda field theories. Such concept is based on a specific ansatz of our basic superfields on a well-defined subspace. We study the integrability criterion of such models by the Lax method.

This paper is structured as follows. In Section 2, we introduce an approach to deriving Liouville equations from the action of the superstring, and we present the supersymmetric formulation and the integrability criterion. We finish this section by studying the case of the gauge string theory. In Section 3, we present our basic notations of the super Toda theory from the 2D superstring worldsheet theory, and we also discuss the integrability of our system. Section 4 is devoted to conclusions and remarks.

2. Toward super-Liouville field theory

2.1 On RNS superstring action

Let us consider the worldsheet action for the superstring theory with a conformal gauge in two-dimensional worldsheet supersymmetry as follows [1, 8].

$$S_0 = -\frac{1}{4\pi\alpha'}\int d^2\sigma\left(\partial^a X^\mu \partial_a X_\mu - i\overline{\psi}^\mu \rho^a \partial_a \psi_\mu\right), \tag{1}$$

with $\sigma^a = (\tau, \sigma)$, $a = 0, 1$, $X^\mu(z, \overline{z})$, $\mu = 0, .., 9$ are spacetime coordinates for a string, ψ^μ are a set of Majorana spinors, and ρ^a are the Dirac matrices written in the Majorana basis as follows

$$\rho^0 = \begin{pmatrix} 0 & -i \\ i & 0 \end{pmatrix}; \quad \rho^1 = \begin{pmatrix} 0 & i \\ i & 0 \end{pmatrix}. \tag{2}$$

In our paper, we consider two types of flat metrics, which are associated with two-dimensional worldsheet surface, and $\eta^{\mu\nu} = diag(-1, 1, \ldots, 1)$ is associated with the ten-dimensional spacetime. However, by using the variation of action (1), we have the following equations of motion

$$\partial_a \partial^a X^\mu = 0, \quad \rho^a \partial_a \psi^\mu = 0, \quad \partial_a \overline{\psi}^\mu \rho^a = 0. \tag{3}$$

The supersymmetric transformations consist of finding the transformations relative to the bosonic and fermionic variables; however, in our case, we have the following transformations

$$\delta X^\mu = \overline{\epsilon}\psi^\mu \quad , \quad \delta\psi^\mu = -i\rho^a \partial_a X^\mu \epsilon \ , \tag{4}$$

where ϵ are infinitesimally small parameters. Using the principle of least action $\delta S = 0$, we obtain the following form of the conserved supercurrent

$$j_a = \frac{1}{2}\rho^b \rho_a \psi^\mu \partial_b X_\mu, \quad \partial^a j_a = 0. \tag{5}$$

In what follows, we will use this formalism to describe the motion of the worldsheet in the form of nonlinear Liouville and Toda equations.

2.2 Liouville's field equations

This section is based on one of our previous works concerning Liouville theory from superstring field theory [18], we adopt the ansatz proposed in such a paper concerning X^μ and ψ^μ, namely

$$X^\mu := \varphi k^\mu, \qquad \psi^\mu := \psi k^\mu, \tag{6}$$

where φ and ψ are two Lorentz scalar fields and k^μ as a constant basis vector that satisfies $k^\mu k_\mu = k^2$, $\mu = 0, 1, \dots, 9$. In accordance with this hypothesis (6), the action of the superstring becomes

$$S_0 = -\frac{k^2}{4\pi\alpha'}\int d^2\sigma\left(\partial^a\varphi\partial_a\varphi - i\overline{\psi}\rho^a\partial_a\psi\right). \tag{7}$$

In order to find the Liouville equations, we have to make a transformation on the action (7), and such a transformation is considered as a perturbation with a weak coupling $\lambda = -\frac{k^2}{4\pi\alpha'}$ that we propose as follows

$$S_0 \to S_0 + \lambda\int d^2\sigma\left(e^{2\varphi} + i\overline{\psi}\psi e^{\varphi}\right). \tag{8}$$

Using standard variational calculus, we can show that the corresponding equations of motion have the following form

$$\partial_a\partial^a\varphi = e^{2\varphi} + \frac{i}{2}\overline{\psi}\psi e^{\varphi}, \quad \rho^a\partial_a\psi = \psi e^{\varphi}, \quad \partial_a\overline{\psi}\rho^a = -\overline{\psi}e^{\varphi}. \tag{9}$$

These equations define the Liouville equations of motion. However, we need to find the supersymmetric version of such equations, that is why we will introduce in the following the superfield formulation of the theory.

2.3 Superfield formulation

By construction, super-Liouville field theory is the supersymmetric generalization of bosonic Liouville theory, which is considered a two-dimensional matter-induced theory of gravity. However, the super-Liouville theory describes 2D supergravity induced by supersymmetric matter [5]. Such a construction is based on the introduction of superfields associated with the supersymmetry $N = 1$ which can be taken as follows [7].

$$\Phi = \varphi + \overline{\theta}\psi + \theta\overline{\psi} + \theta\overline{\theta}F \tag{10}$$

where φ, ψ, and F are the component fields of Φ, and in this case, we can prove that the super-Liouville field equations can be expressed in terms of the supersymmetric derivative D as follows

$$\overline{D}D\Phi = \exp(\Phi) \tag{11}$$

where $D = \partial_\theta + \theta\partial$ and $\overline{D} = \partial_{\overline{\theta}} + \overline{\theta}\partial$. Indeed, straightforward computations lead to

$$\overline{D}D\Phi = -F - \overline{\theta}\partial\overline{\psi} + \theta\partial\psi + \theta\overline{\theta}\partial\overline{\partial}\varphi \tag{12}$$

Performing the exponential expansion of the superfield Φ, we find

$$\exp(\Phi) = \left[1 + \overline{\theta}\psi + \theta\overline{\psi} + \theta\overline{\theta}(-\overline{\psi}\psi - \exp(\varphi))\right]\exp(\varphi) \tag{13}$$

By inserting (12) and (13) in the Eq. (11) and using the complex transformations $z = \sigma + i\tau$ and $\overline{z} = \sigma - i\tau$, we can rewrite the super-Liouville eq. (11) as follows

$$\partial\overline{\partial}\varphi = -\overline{\psi}\psi\exp(\varphi) - \exp(2\varphi); \quad \partial\psi = \overline{\psi}\exp(\varphi); \ \overline{\partial\psi} = -\psi\exp(\varphi) \tag{14}$$

where $\partial_z = \partial$ and $\partial_{\overline{z}} = \overline{\partial}$.

Note that the only constant of motion of this theory is the momentum energy tensor T_2 of conformal weight 2 verifying [19, 20].

$$T_2(z) = \partial^2\varphi - (\partial\varphi)^2, \qquad \overline{\partial}T_2(z) = 0 \tag{15}$$

In what follows, we will study the Lie algebra structures and the integrability criterion of our theory.

2.4 Integratability criterion

As mentioned in [21], the Lie superalgebra associated with the super-Liouville equations is given by the superalgebra $osp(1,2)$ which is formed by three bosonic generators $e^{\pm}$ and h of the Lie algebra $sl(2)$ and of two fermionic generators $f_{\pm}$. The structure of the superalgebra $osp(1,2)$ is given by the following (anti)commutation relations:

$$\begin{gathered}[h, e_{\pm}] = \pm 2e_{\pm}; \quad [h, f_{\pm}] = \pm f_{\pm}; \quad [e_+, e_-] = h; \quad [f_{\pm}, e_{\pm}] = 0 \\ [f_{\pm}, e_{\mp}] = -f_{\mp}, \quad \{f_{\pm}, f_{\pm}\} = \pm i e_{\pm}, \quad \{f_+, f_-\} = \frac{-i}{2}h\end{gathered} \tag{16}$$

in terms of $\mathbb{Z}_2$-graded vector space the superalgebra $osp(1,2)$ is written in the form

$$osp(1,2) = (h, e_+, e_-)_{\overline{0}} \oplus (f_+, f_-)_{\overline{1}} \tag{17}$$

here $\overline{0}$ and $\overline{1}$ represent the bosonic and fermionic generators, respectively.

One of the integrability test methods is given by the existence of the Lax pair [18]. In this system, the determination of the Lax pair helps us to study the integrability criterion of the obtained super-Liouville theory. Indeed, the equation defining the null curvature of our system is given by

$$DA_{\overline{\theta}} + \overline{D}A_{\theta} + \{A_{\theta}, A_{\overline{\theta}}\} = 0 \tag{18}$$

here the generators of the Lax pair $(A_{\theta}, A_{\overline{\theta}})$ are given as functions of the generators of the super Lie algebra $osp(1,2)$. A possibility to do this realization is given by

$$A_{\theta} = f_+ + D\Phi\, h; \ A_{\overline{\theta}} = -2i\exp(\Phi)\, f_- \tag{19}$$

after a simple calculation, we get

$$DA_{\overline{\theta}} = -2iD\Phi \exp(\Phi) f_{-}; \quad \overline{D}A_{\theta} = \overline{D}D\Phi\, h \tag{20}$$

$$\{A_{\theta}, A_{\overline{\theta}}\} = -\exp(\Phi)\, h + 2iD\Phi \exp(\Phi) f_{-} \tag{21}$$

Now, by using the null curvature condition (18) and the commutation relations (16) of the super Lie algebra $osp(1,2)$, we obtain the following super-Liouville equation of motion

$$\overline{D}D\Phi = exp\,\Phi \tag{22}$$

In the next paragraph, we treat the case of superstring field theory in the existence of gauge superfields on the worldsheet superspace.

2.5 Gauged superstring action

The gauged superstring theory has been the subject of several studies [22]. It is obtained by the introduction of abelian gauge fields on the worldsheet. Such fields appear as spacetime coordinates. Let us start with the following bosonic action

$$S_1 = -\int d^2\sigma \left(\frac{1}{4\pi\alpha'} \partial_a X^{\mu} \partial^a X_{\mu} + \frac{1}{4g^2} F_{ab} F^{ab} \right), \tag{23}$$

where g is the gauge coupling constant, $F_{ab} = \partial_a A_b - \partial_b A_a$ is the abelian field strength associated with gauge field A, $a; b = 0, 1$ in the two-dimensional worldsheet surface, and $\mu = 0, .., 9$ in the ten-dimensional spacetime. Therefore, we have the following gauge condition

$$\partial_a A^a = 0. \tag{24}$$

the supersymmetric action is obtained by inserting the following superfields

$$Y^{\mu}(\sigma, \tau, \theta^1, \theta^2) = X^{\mu}(\sigma, \tau) + \overline{\theta}\psi^{\mu}(\sigma, \tau) + \frac{1}{2}\overline{\theta}\theta B^{\mu}(\sigma, \tau), \tag{25}$$

$$\mathcal{A}^a(\sigma, \tau, \theta^1, \theta^2) = A^a(\sigma, \tau) + \overline{\theta}\rho^a \chi(\sigma, \tau) + \frac{1}{2}\overline{\theta}\theta W^a(\sigma, \tau), \tag{26}$$

Here, the fields B^{μ} and W^a represent auxiliary fields, and the field χ is a Majorana-like spinor that represents the supersymmetric partner of A_a. With the two Grassmann variables θ^1 and θ^2, we can form a resultant Majorana spinor variable as follows:

$$\theta = \begin{pmatrix} \theta^1 \\ \theta^2 \end{pmatrix}, \quad \overline{\theta} = \begin{pmatrix} \overline{\theta}^1 \\ \overline{\theta}^2 \end{pmatrix} \tag{27}$$

We must also modify our covariant derivatives by introducing the following super covariant derivatives

$$\mathcal{D}^a = \kappa\varepsilon^{ab}\rho_b\mathfrak{D}, \qquad \mathfrak{D} = \frac{\partial}{\partial\overline{\theta}} - i\rho^a\theta\partial_a, \tag{28}$$

where $\varepsilon^{01} = -\varepsilon^{10} = 1,\ \varepsilon^{00} = \varepsilon^{11} = 0$, and $\kappa \in \left\{\pm\frac{1}{\sqrt{2}}, \pm\frac{i}{\sqrt{2}}\right\}$ is a constant which verifies $\overline{\mathcal{D}}^a Y_\mu \mathcal{D}_a Y^\mu = \overline{\mathfrak{D}} Y_\mu \mathfrak{D} Y^\mu$.

However, such existence of gauge fields implies that the action is transformed into

$$S = -\frac{1}{4\pi\alpha'}\int d^2\sigma\left(\partial_a X^M \partial^a X_M - i\overline{\psi}^M \rho^a \partial_a \psi_M - B^M B_M\right) \tag{29}$$

where X^M, B^M are the bosonic fields of conformal weight, respectively, 1 and 2, ψ^M is the fermionic field of conformal weight $\frac{3}{2}$ and $M \in \{\mu, a\}$ such as

$$\{X^M\} = \{X^\mu\} \cup \{X^a\}, \tag{30}$$

Our hypothesis is to assume that the fields can be written in the following way

$$X^M = \xi^M.\varphi, \quad \psi^M = \xi^M.\psi, \quad B^M = \xi^M.B \tag{31}$$

where ξ^M is the Lorentz constant field of conformal weight 1, and verifying $\xi^M\xi_M = 1$, φ and B are two bosonic fields of conformal weight 0 and 1, respectively, and ψ represents a fermionic field of conformal weight 1/2. Based on the assumption (31), the action (29) takes the following form

$$S_1 = -\frac{k^2}{4\pi\alpha'}\int d^2\sigma\left(\partial_a\varphi\partial^a\varphi - i\overline{\psi}\rho^a\partial_a\psi - B.B\right), \tag{32}$$

the equations of motion of the different fields are given by

$$\partial_a\partial^a\varphi = 0; \quad \rho^a\partial_a\psi = 0; \quad \partial_a\overline{\psi}\rho^a = 0; \quad B = 0. \tag{33}$$

In order to find the Liouville field equations, we will have to consider a specific value for the B field, namely [22].

$$B = e^\varphi + \frac{i}{2}\overline{\psi}\psi. \tag{34}$$

Finally, we find the same equations of motion of Liouville's theory

$$\partial_a\partial^a\varphi = e^{2\varphi} + \frac{i}{2}\overline{\psi}\psi e^\varphi; \quad \rho^a\partial_a\psi = \psi e^\varphi; \quad \partial_a\overline{\psi}\rho^a = -\overline{\psi}e^\varphi \tag{35}$$

The study of the superfield formalism and the integrability criterion is treated in the same way as the previous section.

3. Toward super Toda field theory

Today's theories represent a general framework of Liouville's theory, which are completely integrable classical theories associated with finite-dimensional semi-simple

Lie algebras [23, 24]. Of course, Toda's theory appears in the effective action of two-dimensional string theory with a conforming gauge [25, 26]. However, in the following, we will extend the (super) Toda theory based on a special conformal transformation.

3.1 Bosonic Toda field theory

Let us consider the superstring action with worldsheet super-symmetry which is given by (1). The main goal of our work is to obtain the super-conformal Toda field theory derived from superstring theory related to the action (1). However, as for the ψ^μ fermionic fields, we need the φ scalar fields satisfying the following assumption

$$X^\mu = \frac{k^\mu}{\sqrt{2}}\sum_{i=1}^{n} \mathcal{K}_{ii}\varphi_i; \qquad \psi^\mu = k^\mu \psi \tag{36}$$

the action (1) will now become

$$S_0 = -\frac{k^2}{4\pi\alpha'}\int d^2\sigma \left(\frac{1}{2}\sum_{i,j=1}^{n} \mathcal{R}_{ij}\partial^a\varphi_i\partial_a\varphi_j - i\overline{\psi}\rho^a\partial_a\psi\right) \tag{37}$$

$$\text{with } \mathcal{R}_{ij} = \mathcal{K}_{ii}\mathcal{K}_{jj}, \;\; i,j = 1,..,n \tag{38}$$

where k^μ is a constant vector that satisfies $k^\mu k_\mu = k^2$, $\mu = 0, 1, \ldots, 9$, $\mathcal{K}_{ij}$ is the Cartan matrix of the associated Lie algebra, β represents a coupling constant, and φ_i are scalar fields.

In order to find the super Toda field theory, we take the action (38) and make the following transformation

$$S_0 \rightarrow S_0 + \lambda\int d^2\sigma \left[\frac{1}{\beta}\sum_{i=1}^{n} \exp\left(\beta\sum_{j=1}^{n}\mathcal{R}_{ij}\varphi_j\right)\right] \tag{39}$$

with $\lambda = -\frac{k^2}{4\pi\alpha'}$, and the resulting action is

$$S = -\frac{k^2}{4\pi\alpha'}\int d^2\sigma \left(\frac{1}{2}\sum_{i,j=1}^{n} \mathcal{R}_{ij}\partial^a\varphi_i\partial_a\varphi_j + \frac{1}{\beta}\sum_{i=1}^{n} \exp\left(\beta\sum_{j=1}^{n}\mathcal{R}_{ij}\varphi_j\right) - i\overline{\psi}\rho^a\partial_a\psi\right) \tag{40}$$

By making the action vary with regard to the field φ its derivatives, we find the following equations of motion

$$\partial_a\partial^a\varphi_i = \exp\left(\beta\sum_{j=1}^{n}\mathcal{R}_{ij}\varphi_j\right) \tag{41}$$

However, we have found the classical Toda field equation [18, 20, 27] with the matrix elements $\mathcal{R}_{ij} = \mathcal{K}_{ii}\mathcal{K}_{jj}$.

3.2 Super field analysis

Concerning the superfield formulation associated with the $N = 1$ supersymmetric theory, we can establish the following components

$$\Phi_i = \varphi_i + \overline{\theta}\overline{\psi}_i + \theta\psi_i + \theta\overline{\theta}F_i \; ; i = 1, \ldots \tag{42}$$

where $D = \frac{\partial}{\partial\theta} + \theta\partial$ and $\overline{D} = \frac{\partial}{\partial\overline{\theta}} + \overline{\theta}\partial$. Indeed, straightforward computations lead to

$$D\overline{D}\Phi_i = -F_i - \overline{\theta}\partial\psi_i + \theta\partial\overline{\psi}_i + \theta\overline{\theta}\partial^2\varphi_i \tag{43}$$

expanding the exponential of the superfield Φ, we find

$$D\overline{D}\Phi_i = \exp\left(\beta\sum_{j=1}^{n}\mathcal{R}_{ij}\Phi_j\right) \tag{44}$$

in the case of $osp(3,2)$ Lie algebra, we have this expression for the famous Cartan matrix $\mathcal{K} = \begin{pmatrix} 1 & -1 \\ -1 & 0 \end{pmatrix}$ which implies $\mathcal{R}\begin{pmatrix} 1 & 0 \\ 0 & 0 \end{pmatrix}$, and in this case, the super Toda equation becomes

$$D\overline{D}\Phi_1 = \exp(\beta\Phi_1) \tag{45}$$

our theory admits two conserved currents, $T^{(2)}$ of bosonic character with conformal weight 2 and $T^{(3/2)}$ of fermionic character which plays the role of the supersymmetric partner of $T^{(2)}$ and which has a conformal weight of 3/2. Explicitly we have

$$T^{(3/2)} = \overline{D}^2\Phi_1\overline{D}\Phi_1 - \frac{1}{\beta}\overline{D}^3(\Phi_1 + \Phi_2) \tag{46}$$

$$T^{(2)} = \overline{D}^2\Phi_1\overline{D}\Phi_1\overline{D}\Phi_2 - \frac{1}{\beta}\overline{D}^3\Phi_1\overline{D}\Phi_2 - \frac{1}{2\beta}\left(\overline{D}^2\Phi_2\right)^2 - \frac{1}{\beta^2}\overline{D}^4\Phi_2 \tag{47}$$

Now, the question that arises is this an integrable theory?

3.3 Criterion of integrability

A key step in proving integrability of Toda superstring theory is to establish explicitly the generators of Lax pairs. The null curvature condition is defined as follows

$$DA_{\overline{\theta}} + \overline{D}A_{\theta} + \{A_{\theta}, A_{\overline{\theta}}\} = 0 \tag{48}$$

with the super Toda Lax pair is given by:

$$A_{\overline{\theta}} = \sum_{i=1}^{n}\overline{D}\Phi_i h_i + \sum_{i=1}^{n} e_i; \quad A_{\theta} = \sum_{i=1}^{n} f_i \exp\left(\beta\sum_{j=1}^{n}\mathcal{R}_{ij}\Phi_j\right) \tag{49}$$

where

$$\left[e_i, f_j\right\} = \delta_{ij} h_j, \quad \left[h_i, e_j\right\} = \mathcal{R}_{ij} h_j, \quad \left[h_i, f_j\right\} = -\mathcal{R}_{ij} f_j, \tag{50}$$

By using the commutation relations of the Lie superalgebra $osp(3,2)$ and also the zero curvature condition, we directly obtain

$$D\overline{D}\Phi_i = \sum_{i, fermionic} \exp\left[\beta \sum_j \mathcal{R}_{ij}\Phi_j\right] + \overline{\theta}\theta \sum_{i, bosonic} \exp\left[\beta \sum_j \mathcal{R}_{ij}\Phi_j\right] \tag{51}$$

which are exactly the equations of motion of the super Toda field theory (44). Therefore, we can have the same results of the papers [25, 26, 28] by replacing $\mathcal{K}_{ij}$ by $\mathcal{R}_{ij}$.

4. Conclusion

In this paper, we have discussed in detail the notion of 2D integrable models that live in the worldsheet of the Ramond–Neveu–Schwarz superstring type theory. The resulting nonlinear model can describe the motion or oscillation of this worldsheet in a ten-dimensional spacetime.

First, we recall the action of the superstring theory, its equations of motion as well as its supersymmetric transformations.

However, we studied how we can extract the Liouville equations from the action of 2D worldsheet superstring theory, we expressed the supersymmetric model by introducing the superfield formalism, and we examined the integrability via the Lax equation with the existence of the associated Lax pair. In the case where our theory presents gauge fields on the worldsheet manifold, we have treated this case by taking into account this coupling.

Afterward, we have the opportunity to remake our study for a general case, namely the Toda field theory, and we use the same approach to link this model to the worldsheet of the RNS superstring theory and study the integrability criterion.

Author details

Abderrahman El Boukili*†, Hicham Lekbich† and Najim Mansour†
Equipe de Physique Théorique et Modélisation, Faculté des Sciences et Techniques, Université Moulay Ismail, Errachidia, Morocco

*Address all correspondence to: a.elboukili@umi.ac.ma

† These authors contributed equally.

References

[1] Witten E. On background-independent open-string field theory. Physical Review D. 1992;**46**:5467

[2] Sen A, Zwiebach B. Tachyon condensation in string field theory. JHEP. 2000;**3**:2

[3] Mansour N, Diaf EY, Sedra MB. Electronic Journal of Theoretical Physics. 2018;**14**:21

[4] Mansour N, Diaf EY, Sedra MB. Journal of Physical Studies. 2019;**23**:1103

[5] Neveu A, West PC. Gauge symmetries of the free supersymmetric string field theories. Physics Letters B. 1985;**165**(1–3):63-66

[6] Roy SM, Singh V. Ramond-Neveu-Schwarz string with new boundary conditions. Physics Letters B. 1988; **214**(2):182-186

[7] Polyakov D. Interactions of massless higher spin fields from string theory. Physical Review D. 2010;**82**:066005

[8] Ballestrero A, Maina E. Ramond-Ramond closed string field theory. Physics Letters B. 1986;**182**(3–4):317-320

[9] Damour T, Polyakov AM. String theory and gravity. General Relativity and Gravitation. 1994;**26**:1171-1176

[10] Friedan D, Qiu Z, Shenker S. Superconformal invariance in two dimensions and the tricritical Ising model. Physics Letters B. 1985;**151**(1): 37-43

[11] Rosly AA, Schwarz AS, Voronov AA. Superconformal geometry and string theory. Communications in Mathematical Physics. 1989;**120**:437-450

[12] Susskind L. Dynamical theory of strong interactions. Physics Review. 1967;**154**:1411

[13] Aharonov Y, Komar A, Susskind L. Superluminal behavior, causality, and instability. Physics Review. 1969;**182**:1400

[14] Dijkgraaf R, Verlinde H, Verlinde E. Topological strings in $d < 1$. Nuclear Physics B. 1991;**352**(1):59-86

[15] Ooguri H, Strominger A, Vafa C. Black hole attractors and the topological string. Physical Review D. 2004;**70**: 106007

[16] Mariño M. Chern-Simons theory and topological strings. Reviews of Modern Physics. 2005;**77**:675

[17] Gorsky A, Krichever I, Marshakov A, Mironov A, Morozov A. Integrability and Seiberg-Witten exact solution. Physics Letters B. 1995;**355**(3–4): 466-474

[18] Bilal K, Nach M, El Boukili A, Sedra MB. Advanced Studies in Theoretical Physics. 2011;**5**(2):91-96

[19] Abdalla E, Abdalla MCB, Dalmazi D. And Koji Harada, correlation functions in super Liouville theory. Physical Review Letters. 1992;**68**:1641

[20] Liao HC, Mansfield P. Light-cone quantization of the super-Liouville theory. Nuclear Physics B. 1990;**344**(3): 696-730

[21] Zhang YZ. N-extended super-Liouville theory from $OSP(N, 2)$ WZNW model. Physics Letters B. 1992;**283**(3–4): 237-242

[22] EL Boukili A, Nach M, Bilal K, Sedra MB. International Scholarly

Research Notices. 2013;**2013**:5. Article ID 364852. DOI: 10.1155/2013/364852

[23] Bilal A, Gervais JL. Extended $C = \infty$ conformal systems from classical Toda field theories. Nuclear Physics B. 1989; **314**(3):646-686

[24] Bilal A, Gervais JL. Systematic approach to conformal systems with extended Virasoro symmetries. Physics Letters B. 1988;**206**(3):412-420

[25] Nohara H, Mohri K. Extended superconformal algebra from super Toda field theory. Nuclear Physics B. 1991; **349**:253-276

[26] Nohara H. Extended Superconformal algebra as symmetry of super Toda field theory. Annals of Physics. 1992;**214**:1-52

[27] Ward RS. Discrete Toda field equations. Physics Letters A. 1995;**199**: 45-48

[28] Aliyu MDS, Smolinsky L. Nonlinear Dynamics and Systems Theory. 2005; **5**(4):323-344

Chapter 3

Ultrafast Spectroscopy of Topological Materials

Alexis Chacón

Abstract

Topological materials and topological phases of matter are of unique interest to fundamental sciences and advanced technologies. They exhibit conducting states at the surface or edge and insulating states in the bulk of the material which are protected by symmetries and topological invariants. Ergo, topological materials are robust against perturbations and dissipations. These features are extremely attractive for the development of topological quantum devices. However, the control and diagnosis of topological phases and transitions still are in their early phase and demand intense research. High-order harmonic generation (HHG) produced by the interaction of a strong mid-infrared laser with solids is becoming a unique tool to explore the ultrafast electron dynamics and the electronic structure. In this chapter, we will review the recent theoretical and experimental efforts in studying topological materials *via* the application of attosecond physics tools. We will start with the paradigmatic topological Haldane model and show how the helicity and circular dichroism of the HHG spectra can contain information on the topological phases and transitions of this model. We will review the recent experiments of HHG in 3D topological insulators, how HHG is used to diagnose these materials, and the ultrafast lasers control the topological phases.

Keywords: topological materials, conducting, insulating, symmetries, opological invariants, topological devices, high-order harmonic generation, mid-infrared lasers, helicity, circular dichroism, optical control of topological phases

1. Introduction

In condensed matter physics, the study of microscopic spin and charge transports is currently a subject of intense research, since electronic transport features can lead to revolutionary advanced technologies, that is, efficient and low cost energy transistors, the construction of electronic circuits, and quantum devices [1]. So far, the production of electronic devices has been possible by the control of the transport properties of semiconductors. However, nowadays, the understanding of varios quantum Hall states promises unique opportunities to a technological revolution in the production of new class of materials, that is, topological insulators.

With the experimental discovery of the Integer and Fractional Quantum Hall Effects (IQHE and FQHE) in two-dimensional (2D) materials or electron gas in 1980 and 1982 [2, 3] the study of *topological orders* of matter began. The IQHE occurs when a 2D electron system is subjected to a strong homogenous magnetic field at low temperature and the Hall resistivity ρ_{xy} (or conductivity $\sigma_{xy} = \frac{1}{\rho_{xy}}$) is measured. Turn out that experiments show that this resistivity exhibits different plateaus as a function of magnetic fields—a quantized feature [2]. A quantum mechanics calculation according to Landau's theory demonstrated that this plateau of the ρ_{xy} is quantized in units of h/e^2 and reads $\rho_{xy} = \frac{h}{\nu e^2}$, where h and e are Planck's constant and fundamental electron charge, and ν the filling parameter (in case of the IQHE, ν is a discrete integer number) [4]. The state responsible for IQHE does not break any symmetries. However, it defines a *topological phase* in the sense that there exist fundamental quantities, for instance, the parameter ν and the quantized resistivity, which are invariants concerning smooth variations of the material properties, dissipations, and perturbations [5]. In case of the FQHE, ν is a fractional number for the Hall resistivity ρ_{xy} [3].

1.1 Quantum hall effects and topological Chern insulators

The realization of the IQHE and FQHE leads to the question of whether or not this quantized resistivity (or conductivity) can be experimentally observed without the application of any magnetic field. This will incentivize new research in the production of different materials with the aim to develop quantum devices.

Inspired in this question, Haldane in 1988 proposed that by breaking the Time-Reversal $\mathcal{T}$ Symmetry (TRS) and the Inversion $\mathcal{P}$ Symmetry (IS) of a Tight-Binding Hamiltonian $H(\mathbf{k})$ in "graphene," a quantized Hall conductivity can arise *without any net magnetic flux* through the unit cell of the periodic 2D honeycomb lattice [5, 6]. The keys of Haldane model are to break the $\mathcal{T}$ symmetry by introducing a complex next-nearest-neighbor (NNN) hopping $t_2 e^{i\phi}$, where t_2 is the amplitude of the NNN and ϕ is the *magnetic flux* over the sub-lattice, and simultaneously breaking its $\mathcal{P}$ symmetry by a local on-site potential M_0 in the Hamiltonian of the system [6]. This symmetry breaking led to the quantized Hall resistivity, which was given exactly by $\rho_{xy} = \frac{h}{\nu e^2}$, where ν now is the *topological invariant* or *Chern number* protected by the reflection symmetry. This invariant parameter is defined as the flux of the Berry curvature through the Brillouin Zone (BZ) and quantizes the Hall voltage of the 2D electron system:

$$\nu_m = \frac{1}{2\pi} \int_{\mathrm{BZ}} d\mathbf{k} \cdot \Omega_m(\mathbf{k}). \tag{1}$$

The bulk Berry curvature of the mth-band is $\Omega_m(\mathbf{k}) = \nabla_{\mathbf{k}} \times \xi_m(\mathbf{k})$, where $\xi_m(\mathbf{k}) = i\langle u_{m,\mathbf{k}}|\nabla_{\mathbf{k}}|u_{m,\mathbf{k}}\rangle$ is the Berry connection, and $|u_{m,\mathbf{k}}\rangle$ is the periodic part of the crystal Bloch wavefunction. In the Haldane model, ν is a topological band property and for the "valence" band, it can only be $\nu =$ -1, 0, and + 1 depending on the material properties. Thus, the Haldane model exhibits three different topological phases. In other words, ν is quantized and defines a phase diagram as a function of the Hamiltonian parameters (M_0/t_2 and ϕ) and symmetries [6]. The Chern number ν remains a *constant integer* against smooth changes of the bulk energy band structure, unless a *topological phase transition* takes place. This means that if the material parameters smoothly change (or perturbations are present), also the *bulk band gap varies*. But the gap does not close because the topological invariant ν of $H(\mathbf{k})$ does not change. That invariance character of ν defines ρ_{xy} as a *topological protected and robust state* against electron scattering, electron-electron interactions and dissipations [7]. It is interesting to point out that the *bulk band gap* has to remain open in the system even for different material properties, while the topological Chern number *protects* the quantized Hall conductivity σ_{xy}. Based on the fact that there exists a bulk band gap and a topological quantized conductivity σ_{xy} (or resistivity ρ_{xy}) related to symmetries, these invariant features define a new class of insulator material called *topological Chern insulator*: a 2D material that conducts or transports electrical carriers at the edge but is an insulator at the bulk of the material [5]. Moreover, if the topological invariant changes, the bulk band gap has to pass through a gapless point—known as the critical point. Under that condition, a topological phase transition occurs and the material might exhibit quantized Hall conductance or not. For instance, in case of $\nu = 0$, the topological Chern insulator passes from a topological conducting edge ($\nu = +1$ or -1, $\sigma_{xy} = \pm\frac{e^2}{h}$) state (but insulating bulk state) to an ordinary insulator in which $\sigma_{xy} = 0$ (the bulk and edge do not conduct carriers) [7].

The magnitude $\Omega_m(\mathbf{k})$, Berry curvature, is also a characteristic property of the band structure, which is defined in terms of the quantum mechanics periodic Bloch states $|u_{m,\mathbf{k}}\rangle$. Thus, the bulk Berry curvature is an unique quantity that encodes the quantum mechanics states in the BZ, and is also responsable of various quantum Hall states, anomalous Hall effects, anomalous velocities, valley Hall conductivities, etc. In Eq. (1), the Berry curvature flux also defines the so-called *Berry phase* that can be represented as the closed loop integral in the BZ of the Berry connection $\xi_m(\mathbf{k})$ [5, 7].

Haldane's work promises the hope of a solid-state material that can exhibit a quantized Hall resistivity by breaking the own $\mathcal{T}$ and $\mathcal{P}$ symmetries of the Hamiltonian. The experimental observation of the Haldane Chern insulator has been realized in ultra-cold atoms in a periodically modulated optical honeycomb lattice [8]. However and despite this paradigmatic Haldane prediction of the topological quantum Hall effect with a *zero net magnetic flux*, its experimental demonstration remains challenging for solid-state material sciences. It is only very recently that an experiment claims that a *Haldane Chern insulator* can be observed in an AB-staked $MoTe_2/WSe_2$ [9]. Zhao et al., after applying an external magnetic field in AB-staked $MoTe_2/WSe_2$ moiré bilayers that first exhibit a quantum spin Hall insulator, demonstrate that a Chern number arises in the system. Since an external magnetic field is applied to the unit cell of the moiré bilayers, a total magnetic flux should be different from zero—contradicting the key hypothesis of the Haldane model. Hence, one can ask whether or not this study achieved the experimental realization of the Haldane Chern insulator.

Other questions that also arise are: 1) how the topological invariant or topological order ν can be measured to diagnose the insulating topological Haldane bands? and 2) how can we engineer the NNN hopping $t_2 e^{i\phi}$ by using strong driving lasers and creating topological control of the Haldane phases. We will address these questions in the below sections.

1.2 Topological insulators and quantum spin hall effect

Breaking the time-reversal symmetry of the material is challenging, this is the reason why the experimental observation of the Haldane model has been elusive so far. Kane and Melle were the firsts who proposed a model considering a system which preserves the $\mathcal{T}$

symmetry. In some 2D (or 3D) materials, there are cases where the *Spin-Orbit Coupling* (SOC) becomes strong. That allows different insulating topological bands protected by the $\mathcal{T}$ symmetry of the Hamiltonian [10–12]. These materials belong to a new classification, namely, *topological insulators* (TIs) and are based on the quantum spin Hall effect (QSHE). Kane and Melle in 2005 showed that the key to understanding this QSHE is to examine the role of $\mathcal{T}$ symmetry for spin-1/2 fermions [10, 13]. Here, we will follow their arguments to examine the main features of 2D (or 3D) topological insulators.

Since the Hamiltonian of the 2D honeycomb lattice of Kane-Melle model for graphene is time-reversal, because of the spin-1/2 associated with the NNN hopping parameter, Kramers' degeneracy plays an important role in the structure of the bulk valence and conduction bands. Under the above conditions, Kramers' theorem establishes that each eigen-energy is at least twofold degenerate [5, 7]. In the absence of SOC, the Kramers' degeneracy is simply the degeneracy between up- and down-electron spins. However, in the presence of spin-orbit interaction, Kramers' degeneracy has topologically non-trivial implications. Since the bulk Hamiltonian H has a term regarding with the SOC interaction, $H_{\text{SOC}} = it_2 \sum_{\langle\langle i,j \rangle\rangle, \alpha\beta} \nu_{ij} c_{i\alpha}^{\dagger} s_{\alpha\beta}^{z} c_{j\beta} + h.c.$, where t_2 is the SOC strength between the NNN j and i, $s_{\alpha\beta}^{z}$ is the spin operator (Pauli matrix on z-direction), and $c_{i\alpha}^{\dagger}$ is the electron creation operator with spin index α (spin up and spin down) and $\nu_{ij} = \pm 1$; in the momentum $\mathbf{k}$-space, $\mathscr{H}(\mathbf{k})$ is represented at least by a 4x4 matrix in Kane-Melle approach [13]. The bulk Hamiltonian exhibits a energy band spectra with a gap between the bulk valence and conduction bands about the K/K′ points of the BZ that are twofold degenerate. In addition, since the 2D Hamiltonian is $\mathcal{T}$ invariant, this implies that the total Chern number of the spin valence bands is zero, $\nu = \nu_{\uparrow} + \nu_{\downarrow} = 0$ (ν_s with s standing for up- and down-spins and is the Chern number). Hence, the quantum Hall charge-conductivity is null ($\sigma_{xy} = 0$) [10]. Note, however, in case of specific Hamiltonian parameters with a bulk band gap systems, Kramers' degeneracy allows an additional topological invariant that is $\nu' = 0$ or $\nu' = 1$. The latter defines a new Z_2 invariant and topological classification. This additional *topological order* is a non-trivial feature of the bulk spin-valence bands, which leads to a no-null invariant $\nu' = (\nu_{\uparrow} - \nu_{\downarrow})/2$ defining the *quantum spin Hall state* $\sigma_{xy}^{s} = \frac{e}{2\pi}$— from the accumulated spin-carriers at opposite edges $\mathbf{J}^{s} = \frac{\hbar}{2e}(\mathbf{J}_{\uparrow} - \mathbf{J}_{\downarrow})$ of a 2D material [10, 13]. By the bulk-boundary correspondence, once can infer the existence of *helical edge states* in topological insulators. Physically, this means that the topological edge states allow energy bands with spin-momentum locked properties. Since spin-up and spin-down currents at the edges are protected by $\mathcal{T}$ symmetry, electrons with different spins will propagate with opposite momentum along the edges. This generate two-conducting channels at the edges of the 2D system *protected* by the Z_2 topological invariant. These helical edge states are also robust against smooth deformations of the material, disorders, and perturbations [5].

The Kane-Melle model is an unique proposal for the realization of QSHE. Nevertheless, graphene is not the most suitable material to observe this topological state, since its spin-orbit coupling is low. Alternative, in 2006 Bernevig et al. theoretically suggested a new 2D topological insulator by considering a quantum well of HgTe which has stronger spin-orbit coupling than graphene. Usually, the SOC strength is larger for atomic elements of the periodic table with large atomic number such as Hg. This topological state was based on controlling the quantum well width, d, of HgTe/CaTe that is connected to the Dirac mass term M. Hence, a topological phase transition occurs when M is manipulated *via* d with respect to the critical well thickness

$d_c = 6.3\,\text{nm}$ [14]. For widths $d < d_c$ $(M > 0)$, the quantum well behaves as a standard insulator, but for $d > d_c$ $(M < 0)$ the inverted band structure of the quantum well holds a nontrivial topological state or a *quantum spin Hall effect*. This theoretical proposal was experimentally demonstrated in 2007 by König et al. [11]—showing that a new class of quantum Hall effect can be realized without the application of an external magnetic field. Additionally, 3D topological insulators were proposed in Bi_2Se_3, Bi_2Te_3, and Sb_2Te_3 [15]. The mechanism behind these materials are similar than those for 2D electron systems with QSHE: The strong SOC leads to a bulk band inversion between the valence and the conduction bands, which corresponds to the conducting surface helical states with spin-polarization lock, while the bulk remains gapped and protected by the time-reversal symmetry. However, the calculation of the topological invariant is different and changes in terms of the highly symmetry points of the crystal as a function of the eigen-energy parities [15]. An interesting feature of these 3D topological insulators, in particular, Bi_2Se_3 is that they can be observed at room temperature [5]—making 3D topological states extremely attractive for advanced applications in spintronics and the development of topological quantum transitions. Nevertheless, so far, the measurement and diagnosis of 2D or 3D topological insulators are based on the quantum Hall effects, Angular-Resolved Photoemission Spectroscopy (ARPES), and scanning tunneling microscopy spectroscopy [5]. ARPES provides a direct measurement of the energy dispersion at the surface as well as direct information of the metallic states in 3D topological insulators.

Nevertheless, those techniques require of highly sophisticated experimental setup and the process of diagnosis can damage the topological material or target. Here, a different alternative to extract structural information of topological materials is explored by employing the tools of attosecond sciences applied to topological physics in condensed-matter phases.

1.3 High-order harmonic generation and attosecond physics

The cornerstone tool in attosecond physics is the High-order Harmonic Generation (HHG) phenomenon [16]. HHG is also the based mechanism to produce attosecond pulses of lights—pulses with a time duration in the order of 10^{-18} s= 1.0 *as*—which allow physicists to study the ultrafast electron dynamics in matter [17]. In particular, HHG occurs when an intense ($I_0 \sim 10^{12} - 10^{15}$ W/cm^2) and ultrashort (time-duration $\sim 1.0\,fs = 10^{-15}s$) laser pulse at the infrared (IR, $\hbar\omega_0 = 1.55$ eV) or middle infrared (MIR, $\hbar\omega_0 = 0.2$ eV) domain is focused on a material target, usually a gas, liquid, or solid. The large laser intensity induces a highly nonlinear process in the target that converts the incident photon-energy $\varepsilon_0 = \hbar\omega_0$ into very high multiple integers $\varepsilon_n = n\varepsilon_0$ of the fundamental driving photon energy ε_0 [18, 19] of the laser.

The mechanism behind HHG in gases is based on the three-step model, which was first proposed by Corkum [20] *via* a semiclassical point of view and almost parallel by Lewenstein et al. using a quantum mechanically approach [21]. This mechanism reads as follows: (1) At the maximum of the oscillating laser-field the atomic binding potential is deformed in such a way (forming a potential barrier) that the electron at its ground state is ionized *via* tunneling, then (2) the electron is accelerated by the driving laser and accumulates energy, and (3) once the oscillating electric field changes its sign, the electron comes back to the ion parent and has a probability to return to the ground state. Finally, the electron emits its excess of energy as a form of high photon energy [21].

The resulting HHG spectrum is characterized by three regions: (i) the perturbative region (low harmonic-order range), (ii) the plateau region (inter-medial harmonic-order range), and (iii) the cut-off region (the highest harmonic-order emission) [19]. The cut-off photon-energy is featured by $I_p + 3.2U_p$ where I_p is the atomic ionization potential and $U_p = \frac{e^2E_0^2}{4m\omega_0^2}$, the ponderomotive energy which reads as a function of the electric charge unit e, the electric field strength E_0, the electron mass m, and the frequency of the fundamental oscillating laser field ω_0. We highlight that the HHG cut-off scale quadratically with the electric field strength E_0 and is proportionally inverse to its optical frequency ω_0. Note, in case that the system is inversion symmetry and time-reversal symmetry and a relative long (Full Width Half Maximum (FWHM), FWHM = 200 *fs*, at 800 nm) driving laser is used, the HHG spectrum exhibits only odd harmonic orders.

Since microscopy information regarding the gas target is codified in the high harmonic spectrum, this phenomenon is used to extract electron structural and electron dynamical information from gases [22]. That method is usually called *high harmonic spectroscopy*.

In this chapter, we will review the latest progress of high harmonic spectroscopy in topological materials in order to diagnosis and control the topological phases and transitions by using the features of the high harmonic emission.

2. Solid-state high harmonic generation

Recently, thanks to the optical development of intense middle-infrared (MIR) sources $1.0 - 10.0\,\mu$m, high harmonic generation has been extended to solids and experimentally observed in semiconductors [23, 24] (ZnO or SiO_2). The physical mechanism of HHG in semiconductors and solids is slightly different from that in gases [25, 26]. In solids, (1) the electron is promoted from its valence band to the conduction band *via* tunneling, (2) the electron accumulation of energy is allowed by the shape of both the energy bands and the electron acceleration mediated by the driving laser, and (3) the electron-hole is annihilated in the medium emitting electromagnetic radiation as a form of a high photon-energy.

The experiment shows that the harmonic cut-off scaled linearly with the electric field strength E_0, in contrast to the gas phase [23] that scaled quadratically. This experimental demonstration has generated great attention in Attosecond Sciences and Condensed-Matter Physics communities since the HHG encodes rich information about the electron structure (energy bands) and electron dynamics (Bloch electron oscillations) of the crystalline lattice [27].

In 2017, Stanford's group carries out experiments of high harmonic generation in Transition-Metal Dichalcogenides (TMDCs)—specially, a monolayer of MoS_2 driven by MIR lasers. Considering that the hexagonal lattice structure of MoS_2 breaks the inversion symmetry of the crystalline Hamiltonian, when the linear polarization of the laser field drives the system across the orientation of the crystal axis, the emitted harmonic orders are odd. In contrast and interestingly, the light emission along the perpendicular direction with respect to the laser polarization produces even harmonic orders [28]. The interpretation of these even harmonics was attributed to the breaking of $\mathcal{P}$ symmetry and the appearance of a no null asymmetric Berry curvature $\Omega_m(\mathbf{k})$ but with a Berry curvature flux equal to zero.

Since the features of a topological material are linked to the Berry phase (or Berry curvature), this experiment motivates the question of what physical information carries out the HHG process about the topology of the crystal, in case the Berry curvature flux of the medium is not null within the BZ or in other words, whether or not the harmonic responses of the bulk states and the surface (edges) states of a 3D (or 2D) topological insulator can be used to diagnose the material?

2.1 Topological high harmonics in one-dimensional chains

In 2018, Bauer and Hansen introduced a model of high-order harmonic generation from a 1D periodic lattice similar to the Su-Schrieffer-Heeger (SSH) approximation [29] that exhibits topological features. The structure of this model considers a linear 1D chain of N ions located at the positions x_i, separated by the lattice constant a and with a potential which reads,

$$v_{\mathrm{ion}}(x) = -\sum_{i=1}^{N} \frac{1}{\sqrt{(x-x_i)^2+1}}, \tag{2}$$

$$x_i = \left[i + \frac{N+1}{2}\right]a. \tag{3}$$

Here, the commonly 1D soft-core Coulomb potential for the interaction between ions and electrons is employed, and the atomic units $\hbar = m_e = |e| = 4\pi\epsilon_0 = 1$ are used unless stated otherwise. Bauer considers this potential $v_{\mathrm{ion}}(x)$ in the Density Functional Theory (DFT) and Time-Dependent Density Functional Theory (TDDFT) to calculate a self-consist electronic energy structure ε_i, and the time evolution of the electronic orbitals in order to numerically compute the electrical dipole moment $\langle X(t)\rangle = \sum_{\sigma,i}\int dx x|\psi_{\sigma,i}(x,t)|^2$ as a function of time, once the laser is switched on. Here, $\psi_{\sigma,i}(x,t)$ is the Kohn-Sham orbital calculated via TDDFT, σ stands over the spin index (up $|\uparrow\rangle$ and down $|\downarrow\rangle$ spins), and the index i runs over $1, 2, \ldots, N_\sigma$ where N_σ is the number of Kohn-Sham electrons with spin σ.

In order to treat the appearance of topological edge states, this model undergoes a Peierls transition if the ion positions are modified according to,

$$x_i \to x_{i'} = x_i - (-1)^i\delta,\ i = 1, 2, \ldots, N, \tag{4}$$

the ions are shifted to the left and to the right depending on δ.

This model exhibits three different phases as a function of the parameter $\delta = 0$, $\delta_{\mathrm{A}} > 0$ and $\delta_{\mathrm{B}} < 0$. For $\delta = 0$, the system has equidistant ions with half filled band, that is, a metal. For δ_{A} and δ_{B} the medium behaves as an insulator without edge states and like a topological insulator with edge states, respectively [29].

Figure 1(a) and **(b)** shows the Fourier transform of $\langle X(t)\rangle$ (the HHG spectra) for a full ion dynamics and a frozen ion motions, respectively. The results indicate that there are quantitative differences between the intensity emission from the topological phase B and the topologically trivial phase A of 14 orders of magnitude of the harmonic spectra intensities. This demonstrates clearly the importance of the topological edge states in the process of high harmonic generation suggesting the possibility of associating topological features to the HHG observable. Both harmonic spectra the frozen and the full ion dynamics exhibit the same tendency, indicating that

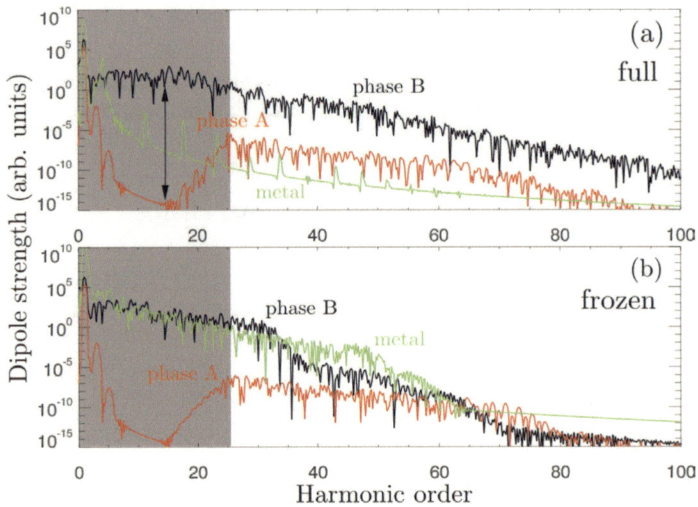

Figure 1.
High harmonic spectra for three different phases: phase A (red line), phase B (black line), and metal phase (green line). (a) And (b) depict the harmonic emissions considering the full variation of the ions in the Kohn-Sham potentials of TDDFT and frozen the ion dynamics, respectively. For both full and frozen configurations, the topological phase (phase B) exhibits an enhancement of 14 orders of magnitude in comparison with the standard insulating phase A—see the black vertical arrow in panel (a). The gray area indicates the region below the band-gap energy of the harmonic emission. The lattice constant is a = 2.0 a.u., the total number of ions is $N = 100$, and the laser parameters where the peak intensity is $I_0 = 2 \times 10^{10}$ W/cm^2, with an optical frequency of $\omega_0 = 0.0075$ a.u. ($\lambda_0 \sim 6.1\mu$m), and total number of optical cycles $N = 5$ opt. Cycles under a $\sin^2$-*envelope.*

topological HHG emission is robust against the ion motions. That fact is important because it highlights a unique feature of a topological observable—its robustness.

This 1D model exhibits interesting properties of topological insulators codified in the HHG spectrum. However, usually, the quantum spin Hall effect or quantum Hall effect (fundamental basis of topological materials) is produced in 2D mediums, which implies a 1D approach is not suitable to describe the topological transversal conductivity. In the next, we will review how to distinguish topological phases and transitions by using high harmonic generation in the paradigmatic Haldane model.

2.2 Haldane model, topological phases, and transitions

By using high harmonic generation resulting from the illumination of MIR laser in topological materials, two different approaches to identify topology and particularly topological phases and transitions were proposed in the paradigmatic Haldane model as shown in Refs. [30, 31]. One of these approaches is the helicity of the harmonic emission and the other one the circular dichroism. We will review the Haldane model and apply it to generate harmonic orders driving by ultrafast lasers in the MIR regime to figure out how the HHG process can capture topological information by both the helicity and the circular dichroism in the HHG spectra.

The tight-binding approximation in the **k**-space leads to the Hamiltonian $H_0(\mathbf{k})$ of Haldane model which is given by,

$$H_0(\mathbf{k}) = B_0(\mathbf{k})\sigma_0 + \mathbf{B}(\mathbf{k}) \cdot \sigma, \tag{5}$$

where the set of $B_0(\mathbf{k})$ and $\mathbf{B}(\mathbf{k}) = \{B_1(\mathbf{k}), B_2(\mathbf{k}), B_3(\mathbf{k})\}$ is known as a pseudo-magnetic field of the tight-binding components, $B_0(\mathbf{k}) = 2t_2\cos(\phi_0)\sum_{i=1}^{3}\cos(\mathbf{k}\cdot\mathbf{b}_i)$; $B_1(\mathbf{k}) = t_1\sum_{i=1}^{3}\cos(\mathbf{k}\cdot\mathbf{a}_i)$; $B_2(\mathbf{k}) = t_1\sum_{i=1}^{3}\sin(\mathbf{k}\cdot\mathbf{a}_i)$; $B_3(\mathbf{k}) = M_0 - 2t_2\sin(\phi_0)\sum_{i=1}^{3}\sin(\mathbf{k}\cdot\mathbf{b}_i)$. Here, $\mathbf{a}_i$ are the Nearest Neighbor (NN) vectors (from atom A to B, see **Figure 2(c)**) of the hexagonal lattice, $\mathbf{a}_1 = (0, a_0)$; $\mathbf{a}_2 = -\frac{a_0}{2}(\sqrt{3}, 1)$; and $\mathbf{a}_3 = \frac{a_0}{2}(\sqrt{3}, -1)$, and the $\mathbf{b}_i$ are the NNN vectors (from atoms A to A or B to B, see **Figure 2(c)**) given by $\mathbf{b}_1 = a_0(\sqrt{3}, 0)$; $\mathbf{b}_2 = \frac{a_0}{2}(-\sqrt{3}, 3)$ and $\mathbf{b}_3 = -\frac{a_0}{2}(\sqrt{3}, 3)$, with a_0 denoting the distance between atomic location A and the atom B. σ_0 is a 2×2 unitary matrix, and σ_i are the Pauli matrices. The symmetries of the model are obtained as follows: (i) A nonzero magnetic flux ϕ_0 breaks the $\mathcal{T}$ symmetry and (ii) a nonzero staggering potential M_0 breaks the $\mathcal{P}$ symmetry.

The calculation of the topological invariant in Haldane model (the Chern number C_m) is mediated *via* Berry's curvature ($\Omega_m(\mathbf{k})$) flux in the BZ. Thus, one requires to compute the Berry connection $\xi_m(\mathbf{k})$ of the m-band, which is defined in terms of the eigenstates of Haldane's Hamiltonian (5). The eigen-energies of Haldane model are given by $\varepsilon_{c/v}(\mathbf{k}) = B_0(\mathbf{k}) \pm \sqrt{B_1(\mathbf{k})^2 + B_2(\mathbf{k})^2 + B_3(\mathbf{k})^2}$ the $+/-$ sign labels the conduction c/valence v band. The Bloch eigenstates of $H_0(\mathbf{k})$ for the conduction and

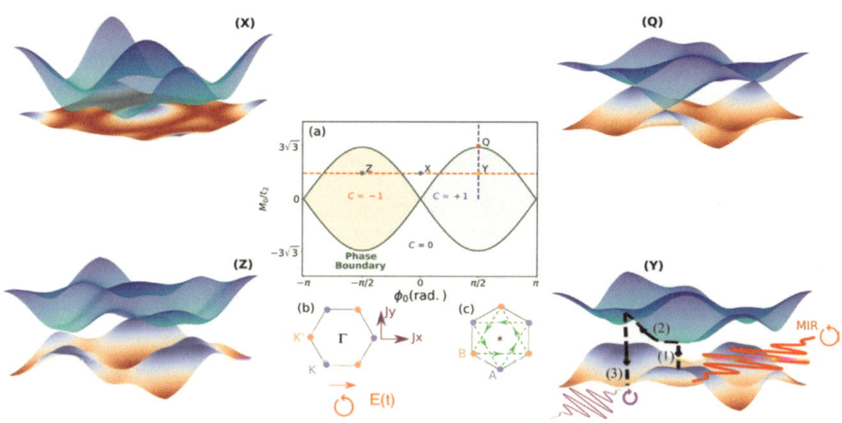

Figure 2.
Topological phase diagram and band structure of the Haldane model: (a) phase diagram of the Haldane model for Chern insulators, the Chern number ($C_v = \nu_v$) for the valence band is depicted as a function of the staggering potential M_0/t_2 (normalized with respect to the second hopping parameter amplitude t_2), and magnetic flux ϕ_0—the index v indicates the valence band label. The green lines exhibit the boundary of the topological phase transition: The topological Chern number changes from $C_v = -1$, 0, to $+1$. The red dashed horizontal line shows a cut of the phase diagram in which topological phase transitions occur, while the inversion symmetry breaking parameter M_0 is kept constant but the time-reversal symmetry parameter ϕ_0 varies. The blue dashed vertical line also shows a topological phase transition considering a fixed magnetic flux $\phi_0 = \frac{\pi}{2}$ and varying the local onsite potential M_0/t_2. (b) And (c) show the honeycomb lattice in the reciprocal and the real spaces, respectively, for the Haldane model. The energy dispersion of the topological bands for different points of the phase diagram (X), (Z), (Q), and (Y) are depicted in surface colors. The (Y) and (Z) points exhibit two distinguished nontrivial topologically band structures with different topological invariants $C_v = +1$ and $C_v = -1$, respectively. For (X) point, the band structure represents a topologically trivial material (a standard semiconductor)—note that there exists a band gap in the energy dispersions for (X), (Y), and (Z) points of the phase diagram. For (Q), the band structure is located at the critical point of the phase diagram in which the bulk band gap is obviously closed at K′ point in the BZ. This clearly suggests that in a topological phase transition the band gap closes and is then reopened. In panel (Y), the high-order harmonic generation mechanism is shown (see text for details).

valence band are defined as: $|u_{+,\mathbf{k}}\rangle = \frac{1}{N_{+,\mathbf{k}}}\{B_3(\mathbf{k})+|\mathbf{B}(\mathbf{k})|, B_1(\mathbf{k}) + iB_2(\mathbf{k})\}$ and $|u_{-,\mathbf{k}}\rangle = \frac{1}{N_{-,\mathbf{k}}}\{iB_2(\mathbf{k}) - B_1(\mathbf{k}), B_3(\mathbf{k})+|\mathbf{B}(\mathbf{k})|\}$, respectively. Here, $N_{\pm}(\mathbf{k})$ is a normalization constant for the eigenstates $|u_{\pm,\mathbf{k}}\rangle$.

Given the above conditions, **Figure 2** shows the calculation of the Chern number as a function of the key ingredients in Haldane model to exhibit topological phases: the inversion symmetry breaking M_0/t_2 and time-reversal symmetry breaking ϕ_0 parameters. The three distinguished topological phases and topological phase transitions are shown in red (C_v=−1), blue (C_v=+1), and white (C_v=0) shadow colors of **Figure 2(a)**. The other panels (X), (Y), (Z), and (Q) show the topological trivial and nontrivial band structure and the critical band at the point in which the topological phase transition is taking place.

2.2.1 High-order harmonics in Haldane Chern insulators

We wish to prove that the light emission from a topological Chern insulator driven by mid-infrared laser can encode the topological features of the band structure and its topological invariant. In order to demonstrate that hypothesis, we require to calculate the high harmonic emission from the paradigmatic Haldane model similar that those done by Bauer and Hansen in a 1D lattice (this will be done by Fourier analyzing the microscopy current). However, instead of solving the TDDFT and calculating the dipole moment, we numerically solve the semiconductor Bloch equations (SBEs) [31] or the Reduced Density Matrix (RDM) [30] and compute the density current as a function of time, to subsequent Fourier transform the currents. That will yield the HHG spectrum. First, we define the microscopy density current as: $\mathbf{J}(t) = \mathbf{J}_{ra}(t) + \mathbf{J}_{er}(t)$ in terms of the intraband current $\mathbf{J}_{ra}(t)$ and the interband current $\mathbf{J}_{er}(t)$ which reads:

$$\mathbf{J}_{er}(t) = e\frac{d}{dt}\mathbf{P}_{er}(t),$$
$$\mathbf{J}_{er}(t) = e\frac{d}{dt}\int_{\overline{\mathrm{BZ}}}\mathbf{d}^{*}_{cv}(\mathbf{K}+\mathbf{A}(t))\rho_{vc}(\mathbf{K},t)d^2\mathbf{K} + c.c. \tag{6}$$

Where $\mathbf{P}_{er}(t)$ is the interband polarization of the medium, $\mathbf{d}_{cv}(\mathbf{k})$=$i\langle u_c(\mathbf{k})|\nabla_{\mathbf{k}}|u_v(\mathbf{k})\rangle$ is the *transition dipole matrix element* between the valence and conduction bands, and $\rho_{vc}(\mathbf{k},t)$ is the coherence of the occupations. In addition, the crystal quasimomentum $\mathbf{K} = \mathbf{k} - \mathbf{A}(t)$ is defined within a moving frame in the Brillouin Zone $\overline{\mathrm{BZ}} = \mathrm{BZ} - \mathbf{A}(t)$ and $\mathbf{A}(t)$=$-\int^t\mathbf{E}(t')dt'$ the vector potential of the driving laser field $\mathbf{E}(t)$. The intraband current reads as follows:

$$\mathbf{J}_{ra}(t) = e\sum_m\int_{\overline{\mathrm{BZ}}}\mathbf{v}_m(\mathbf{K}+\mathbf{A}(t))\rho_{mm}(\mathbf{K},t)d^2\mathbf{K}. \tag{7}$$

Here, $\mathbf{v}_m(\mathbf{k})$=$\nabla_{\mathbf{k}}\varepsilon_m(\mathbf{k}) - \mathbf{E}(t)\times\Omega_m(\mathbf{k})$ is the electronic velocity in the medium as a function of the group velocity $\nabla_{\mathbf{k}}\varepsilon_m$ and anomalous velocity $-\mathbf{E}(t)\times\Omega_m(\mathbf{k})$ components for the m-band. $\rho_m m(\mathbf{K},t)$ The occupation of the m-band. To compute the total density current once requires the knowledge of the density matrix operator $\hat{\rho}(\mathbf{k},t)$ or the occupations ($\rho_{mm}(\mathbf{k},t)$) and coherences ($\rho_{mm'}(\mathbf{k},t)$ $m \neq m'$) of the bands that are governed by the quantum mechanical SBEs:

$$\begin{aligned}\partial_t \rho_{m'm}(\mathbf{K},t) = i\left[\Delta\varepsilon_{m'm}(\mathbf{K}+\mathbf{A}(t)) + \mathbf{E}(t)\cdot\Delta\xi_{m'm}(\mathbf{K},t) + \frac{i}{T_2}\right]\rho_{m'm}(\mathbf{K},t) \\ +i\sum_{m''\neq m'} \mathbf{E}(t)\cdot \mathbf{d}^*_{m'm''}(\mathbf{K}+\mathbf{A}(t))\rho_{m''m}(\mathbf{K},t) \\ -i\sum_{m''\neq m} \mathbf{E}(t)\cdot \mathbf{d}_{mm''}(\mathbf{K}+\mathbf{A}(t))\rho_{m'm''}(\mathbf{K},t).\end{aligned} \tag{8}$$

And,

$$\partial_t \rho_{mm}(\mathbf{K},t) = -\mathcal{I}m\left\{\sum_{m''\neq m} \mathbf{E}(t)\cdot \mathbf{d}_{mm''}(\mathbf{K},t), \rho_{m''m}(\mathbf{K},t)\right\} \tag{9}$$

where $\Delta\varepsilon_{m'm} = \varepsilon_{m'} - \varepsilon_m$ is the energy dispersion difference between m' and m bands, $\Delta\xi_{m'm} = \xi_{m'm'} - \xi_{mm}$ is the Berry connection difference. By solving numerically Eqs. (8) and (9) for the properties of Haldane model and computing the Fourier transform of the microscopy currents (6) and (7), one can obtain the harmonic spectrum for different topological phases of Haldane phase diagram of **Figure 2(a)**.

The calculation of the HHG spectra for different topological phases and laser configurations is shown in **Figure 3**.

The harmonics produced by a linear polarized driving laser from trivial and topological phases are similar in Haldane model. We find a small difference in the intensity emission, but this is not a clear signature that a topological feature will be encoded in a potential experiment (we will not focus on the high harmonic emission driving by linearly polarized lasers). Note, however, similar theoretical results were obtained by Silva et al. [30]. In that paper, instead of using the harmonic emission intensity, they investigated the connection between the harmonic helicity and the topological phases and transitions of Haldane model. A remarkable difference of the helicity as a function of Haldane phase diagram parameters was predicted, indicating that the HHG signal is capable to map indirectly the topological phase transitions.

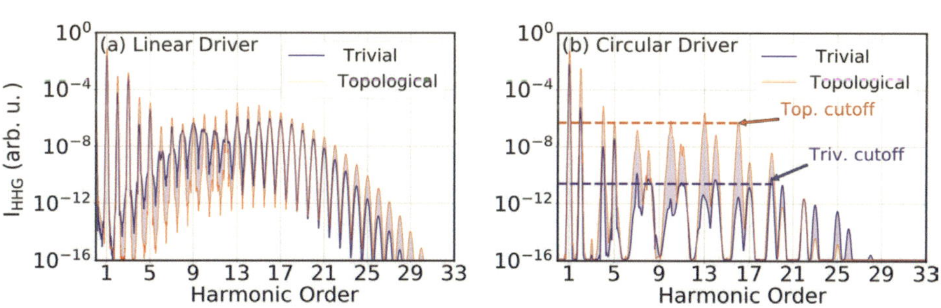

Figure 3.
High harmonic generation spectra from topologically trivial and nontrivial materials. The light intensity emission $I_{HHG}(\omega)$ as function of the harmonic order (ω/ω_0: where ω is the frequency axis and ω_0 the laser frequency) from trivial ($C_v = 0$) and topological ($C_v = +1$) phases driven by lasers with (a) linear polarization along Γ-K direction and (b) right-circular polarization. The violet area points out the intensity emission difference between the trivial and the topological phases. The parameters of Haldane model are: $t_1 = 2.04$ eV, $t_2 = t_1/3$, and $M_0 = 2.54t_2$, with a lattice constant $a_0 = 1.0$°A. The laser is centered at a wavelength of 3.5μm ($\omega_0 = 0.014$ a.u.), which has a Gaussian envelope with 14 opt. Cycles at FWHM and a peak electric-field strength of $E_0 = 0.0045$ a.u. ($I_0 = 7.0 \times 10^{11}$ W/cm^2). We use a dephasing time $T_2 = 5.2$ fs. The magnetic flux of the trivial and topological phases is $\phi_0 = 0.06$ and 1.16 rad., respectively. The band gap is $E_g \sim 3.0$ eV. In (b), the horizontal red dashed line indicates the intensity of the harmonic cutoff for the topological phase while the blue line, for the trivial phase. There is an intensity difference of four orders of magnitude between the topological and trivial phases.

An alternative to explore topological features using high harmonic generation is to produce the harmonic emission by circularly polarized laser (CPL). In **Figure 3(b)** we show HHG spectra from trivial and topological phases driven by a right-hand CPL. We observe three remarkable features: (1) a large enhancement of the harmonic intensity for the topological phase with respect to the trivial one about four orders of magnitude across the harmonics of the plateau (middle region of the HHG spectrum); (2) a shorter harmonic cutoff (the maximum emitted harmonic order) for the topological phase in comparison with the trivial one; and (3) an asymmetric yield between the intensity of the co-rotating harmonic orders ($3n+1$) and the contra-rotating orders ($3n+2$) along the plateau of the HHG spectrum—this asymmetric yield is different for the topological phase and trivial one and can be used as a signature to prove topological phase transitions.

2.2.2 Topological phase transitions with circular dichroism using HHG

So far, we have analyzed the high-order harmonic emission from trivial and topological phases driven by linearly and circularly polarized lasers. We find theoretical evidences that suggest that the harmonic emission produced by circularly polarized laser fields can codify topological features of the Haldane model. In this subsection, we study how the asymmetry between the harmonic yield generated by left-handed and right-handed circularly polarized light; namely, *circular dichroism* (CD) maps the topological phase transitions in Haldane model. Formally, the CD is defined by:

$$\mathrm{CD}_k = \frac{I^{(+)}_{k,RCP} - I^{(-)}_{k,LCP}}{I^{(+)}_{k,RCP} + I^{(-)}_{k,LCP}}, \tag{10}$$

the normalized difference in the harmonic emission driven by left- and right-handed circularly polarized lasers (LCP/RCP). Here, $k = 3n \pm 1$ denote the co-rotating and contra-rotating harmonic orders. The circular dichroism is defined in terms of the circular components of the light emission, $I^{(\pm)}_{k,v} = (k\omega_0)^2 \left| J_{x,v}(k\omega_0) \pm iJ_{y,v}(k\omega_0) \right|^2$, $J_{x,v}$ and $J_{y,v}$ are the x and y components of the spectral density current driven by a v = RCP, LCP laser fields.

A test of the CD hypothesis is to compute the HHG spectra for trivial and topological phases driven by LCP and RCP, and compare the intensity yield of the harmonic orders along the plateau of the spectrum for co-rotating or contra-rotating orders. This test is shown in **Figure 4(a)** and **(b)** for trivial and topological phases, respectively. The asymmetry predicted by Eq. (10) is remarkable different for trivial phase in comparison with the topological phase: The sign of the CD for trivial phase is opposed that the sign for the topological phase. We can then ask whether or not the CD will be sensitive to a topological phase transition across two different lines of Haldane phase diagram shown explicitly in **Figure 2(a)**. So, the circular dichroism as a function of the magnetic flux and local potential is calculated for the harmonic orders around the plateau and is depicted in **Figure 4(c)** and **(d)**, respectively. Clearly, we note that the CD is constant across the topological region in which the Chern number is $C_v = C = \mp 1$; in other words, the CD is negative -1 and positive $+1$ for the corresponding topological phases with $C = -1$ and $+1$. The CD for the trivial phase $C = 0$ is not stable. Ideally, the CD should be zero as the Chern number is null for a trivial

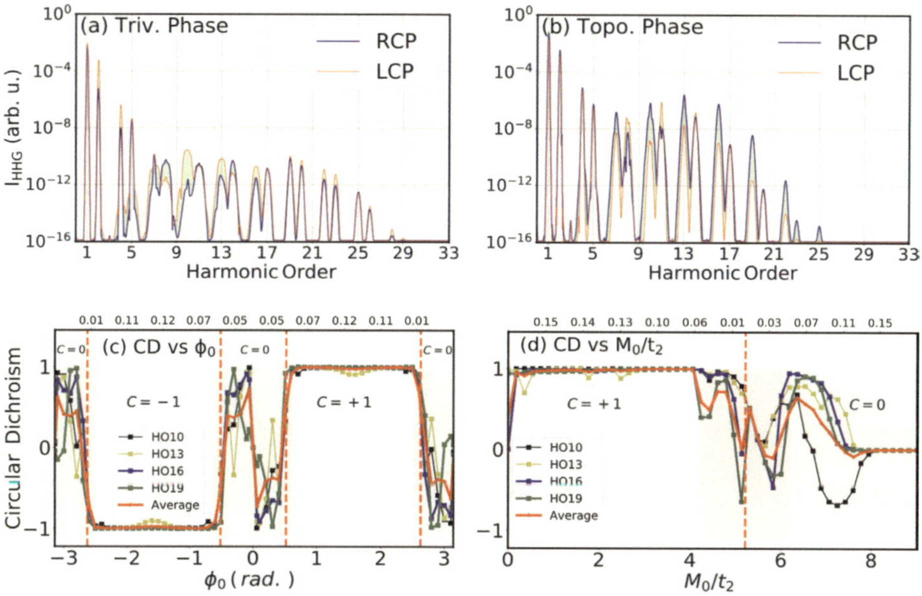

Figure 4.
Circular dichroism for topological phase transitions using HHG process. (a) And (b) depict the high harmonic spectra at equals band gap E_g for trivial and topological phases, respectively, driven by right- and left-circularly polarized lasers with the difference between the two emissions shaded in green (for these simulations, we use the same parameters that in ***Figure 3****). (c) And (d) exhibit the circular dichroism (CD) as a function of the magnetic flux ϕ_o and the staggering potential M_o/t_2 parameters (the scan is defined across dashed lines of* ***Figure 2(a)****) for the plateau harmonic. The top horizontal axis shows the band gap E_g in atomic units (a.u.). The trivial and topological regions are shaded in green and blue, with the critical region (topological critical point in which the phase transition occurs) shaded in red.*

phase; nevertheless, it is "chaotic" specially in **Figure 4(c)**. Note, however, in **Figure 4(d)**, the CD trends to zero right after the critical region.

2.3 Optical nonlinear response in three-dimensional topological insulators

The above theoretical predictions suggest that the physical process of high harmonic generation can encode topological structural information by analyzing the intensity of the harmonic emission, and to prove topological phase transitions by employing the helicity or the circular dichroism observables. Nevertheless, these tests are demonstrated for Chern insulators, which are challenging to produce experimentally in condensed states.

Here, we address the question of whether or not high harmonic generation can be used in a real 3D topological insulator (TI) such as Bi_2Se_3. This prototype TI exhibits topological surface spin-polarized current protected by time-reversal symmetry and the bulk does not conduct carriers or spin-charge. We firstly review the theoretical prediction introduced by Baykusheva et al. in Ref. [32]. This work suggests that the yield of the harmonic intensity as a function of the driven laser ellipticity exhibits an anomalous behavior, which comes from the emission of the topological surface states. Secondly, we will address the experimental verification of this theoretical prediction [33]. In addition, we breafly review the first two experimental realizations of high harmonic emission from other types of 3D-TIs: $BiSbTeSe_2$ and Bi_2Te_3.

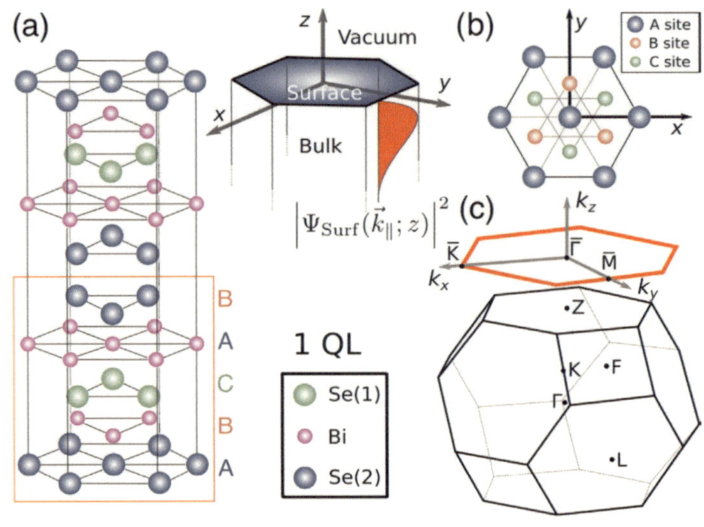

Figure 5.
Crystalline structure of a three-dimensional (3D) topological insulator. (a) Crystal structure of Bi_2Se_3 localizes Bi and Se layers, stacked along the z-direction. Five consecutive layers form a quintuple layer (QL, see red rectangle), the building block of the lattice. Each QL contains five atoms: two equivalent Bi sites, two equivalent Se sites [Se(1)], and a third Se atom, Se(2), which assumes the role of an inversion center, the Hamiltonian of the system preserves the inversion symmetry. The lattice constants are a = 4.14̊ A and c = 28.70̊ A. The space-dependent electron density of the topological surface states, $|\Psi_{\text{Surf}}(\mathbf{k}_{\parallel}, z)|^2$, is depicted in read shaded surface and illustrates the employed boundary conditions—bulk boundary correspondence. (b) Schematic representation of the C_{3v}-symmetric Bi_2Se_3 (111) surface (rhombohedral convention), exposing a top Se layer and underlying Bi and Se′ layers. (c) Sketch of the 3D BZ of bulk Bi_2Se_3 (black) with the four time-reversal-invariant points indicated (Γ, L, F, and Z). The projected 2D BZ of the (111) surface is shown as a red hexagon, with labeled high-symmetry points $\bar{\Gamma}$,$\bar{K}$,$\bar{M}$.

The crystalline atomic structure of the paradigmatic 3D TI Bi_2Se_3 is described by a rhombohedral geometry with five atoms located in five layers as shown in **Figure 5 (a)**, together with its corresponding 3D BZ and the projected 2D surface BZ.

These are stacked along the z-direction and held together by weak van der Waals interactions. The spatial symmetries of the rhombohedral point group include (i) inversion symmetry, (ii) twofold rotation along the x-direction $\hat{R}_2^{(x)}$, (iii) threefold rotation around the x-direction, $R_3^{(z)}$. Additionally, the electronic wave functions or states are featured by the $\mathcal{T}$ symmetry similar to Kane-Mele model. The inversion symmetry (for the (111)-surface of Bi_2Se_3 depicted in **Figure 5(b)**) is necessary lost at the boundary, while the threefold symmetry $R_3^{(z)}$ is preserved as well as a mirror plane $\hat{\sigma}_{\text{reff}}^{(y)}$ coincident with the y-direction. The tight-binding model (TBM) is constructed from the four levels closest to the Fermi level which forms the basis for each site:

$$\left\{|P_z^{(+)}, \uparrow\rangle, |P_z^{(-)}, \uparrow\rangle, |P_z^{(+)}, \downarrow\rangle, |P_z^{(-)}, \downarrow\rangle\right\}. \tag{11}$$

The superscripts $\pm$ denote the parity of the state, and $|P_z^{(+)}, \uparrow(\downarrow)\rangle$ and $|P_z^{(-)}, \uparrow(\downarrow)\rangle$ are derived from p_z orbitals of the Bi and Se atoms, respectively. In the basis defined in Eq. (11), TBM Hamiltonian in the **k**-space has the generic form:

$$\tilde{H}(\mathbf{k}) = \tilde{\varepsilon}(\mathbf{k}) + \sum_{i=1}^{3}\left(\tilde{t}_{a_i}e^{i\mathbf{k}\cdot\mathbf{a}_i} + \tilde{t}_{b_i}e^{i\mathbf{k}\cdot\mathbf{b}_i} + c.c.\right), \tag{12}$$

where $\tilde{\varepsilon}(\mathbf{k})$ is a diagonal on-site energy term. The vectors $\mathbf{a}_i$ and $\mathbf{b}_i$ are the positions of the six intra- and interlayer neighbors on each lattice site (for details, see Ref. [32]). $\tilde{t}_{\mathbf{a}_i}$ and $\tilde{t}_{\mathbf{b}_i}$ denote the intra- and interlayer nearest neighbor hopping parameters. This Hamiltonian can be expanded and written as:

$$\tilde{H}(\mathbf{k}) = h_0(\mathbf{k}) + \sum_{i=1}^{5} h_i(\mathbf{k})\Gamma_i, \tag{13}$$

where Γ_i are the Dirac matrixes defined in terms of Pauli matrices σ_i and τ_i. The auxiliary functions $h_i(\mathbf{k})$ are defined in Appendix A. This Hamiltonian defines whole TBM for the bulk of Bi_2Se_3.

2.3.1 Bulk states

To compute the topological bulk states (TBSs) in Bi_2Se_3, we apply an unitary transformation $\hat{U}_1$ introduced by Liu et al. [34] to the Hamiltonian (13) $\hat{H}(\mathbf{k}) = \hat{U}_1\tilde{H}(\mathbf{k})\hat{U}_1^T$, by diagonalizing the resulting Hamiltonian one can find the eigenenergies of the bulk. The spectrum is doubly degenerated as a consequence of a combined action of IS, TRS, and the spin-orbit coupling. These energy dispersions for the valence band (−) and the conduction band (+) read:

$$\mathcal{E}_{\mathrm{BS}}^{\pm}(\mathbf{k}) = h_0(\mathbf{k}) \pm \sqrt{\sum_{i=1}^{5} h_i^2(\mathbf{k})}. \tag{14}$$

The eigenstates are doubly degenerated as well and the two spinors, labeled by $\nu = \{1, 2\}$, have the form:

$$|\Psi_{\mathrm{B},\nu=1}^{\pm}(\mathbf{k})\rangle = \mathcal{N}_{\mathrm{B}}^{\pm}(\mathbf{k}) f_1(\mathbf{k}) \begin{pmatrix} -i\left(h_5(\mathbf{k}) \pm \sqrt{\sum_{i=1}^{5} h_i^2(\mathbf{k})}\right) \\ -(h_3(\mathbf{k}) + ih_4(\mathbf{k})) \\ 0 \\ h_1(\mathbf{k}) + ih_2(\mathbf{k}) \end{pmatrix}, \tag{15}$$

and

$$|\Psi_{\mathrm{B},\nu=2}^{\pm}(\mathbf{k})\rangle = \mathcal{N}_{\mathrm{B}}^{\pm}(\mathbf{k}) f_2(\mathbf{k}) \begin{pmatrix} 0 \\ h_1(\mathbf{k}) - h_2(\mathbf{k}) \\ i\left(h_5(\mathbf{k}) \pm \sqrt{\sum_{i=1}^{5} h_i^2(\mathbf{k})}\right) \\ h_3(\mathbf{k}) - ih_4(\mathbf{k}) \end{pmatrix}. \tag{16}$$

Here $\mathcal{N}_{\mathrm{B}}^{\pm}(\mathbf{k})$ is a normalization constant and $f_\nu(\mathbf{k}) = \left(h_1 + i(-1)^{\nu-1}h_2(\mathbf{k})\right)^{-1}$. By definition, the spinors $|\Psi_{\mathrm{B},\nu=1,2}^{\pm}(\mathbf{k})\rangle$ form a Kramers doublet, for instants they are related by the time-reversal operator $\hat{\mathcal{T}} = i(\sigma_y \otimes \hat{I}_2)\mathcal{K}$ with $\mathcal{K}$ being the complex conjugate operator. TRS has profound implications for the physics of topological

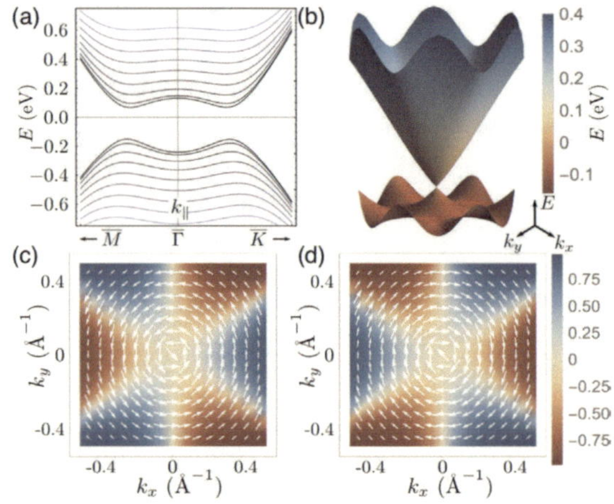

Figure 6.
Topological 3D band structure of Bi_2Se_3 and spin-polarization of the topological surface states. (a) Energy dispersion of the bulk states $\mathscr{E}_B^{\pm}(k_x, k_y, k_z)$ along the high-symmetry directions $\overline{\Gamma M}$ and $\overline{\Gamma K}$ shown for different values of k_z. The black lines correspond to the plane $k_z = 0$, and projected band dispersion curves pertaining to increasing k_z are given in progressively lighter blue colors, whereby the k_z is varied by $\Delta k_z = 1.5 \times 10^{-3}\,\text{Å}^{-1}$. (b) Energy dispersion of the surface states $E_{2D}^{S;(\pm)}(k_x, k_y)$ resulting from the employed TBM model, given in the disk defined by $k_\parallel \leq 0.4\,\text{Å}^{-1}$. (c) And (d) spin polarization of the surface for lower and upper Dirac cones, respectively, over a selected portion of the BZ. The white arrows indicate the magnitude and direction of the in-plane polarization, whereas the color coding corresponds to the magnitude of the spin polarization in z-direction (out of plane). The Berry curvature in momentum space follows a similar pattern as the out-of-plane spin polarization.

insulators. By virtue of the Kramers theorem, no time-reversal-invariant perturbation induces gap opening at the surface Dirac cone. The parameters used in this chapter are extracted from the Table I reported in Ref. [33].

Figure 6(a) shows the resulting bulk band structure $\mathscr{E}_B^{\pm}(\mathbf{k})$ along the parallel momentum $\mathbf{k}_\parallel = (k_x, k_y)$ across the line $\overline{\Gamma M}$ and $\overline{\Gamma K}$ in the BZ for different values of k_z. For simplicity, the interlayer spin-flip hopping parameter is neglected. We appreciate a band gap between the valence and the conduction band around $E_g = 0.37$ eV for the topological bulk energy bands, which is close to the experimentally reported 0.3 eV. The topological states (15) and (16) are used to compute the physical quantities needed to integrated the Semiconductor Bloch Equations or the time-dependent density matrix used to evaluate the charge currents.

2.3.2 Surface states

In order to describe the topological surface states (TSS) of Bi_2Se_3, we first derive an effective two band Hamiltonian $H_{2D}^S(\mathbf{k}_\parallel)$ based on the generic 4×4 TBM Hamiltonian of the topological bulk $H(\mathbf{k})$. Here, we restrict the discussion to a brief recapitulation of the main steps to build the topological surface Hamiltonian. The starting point in our ansatz is to impose open boundary conditions onto the Hamiltonian $H(\mathbf{k})$ by postulating that the surface-state $|\Psi_{\text{Surf}}(\mathbf{k}_\parallel; z)\rangle$ vanishes at the crystal-vacuum interface (defined as $z = 0$ in **Figure 5**) and decays exponentially into the bulk for $z \to \infty$, as illustrated by the red-shaded surface in **Figure 5(a)**. With the aid to apply this procedure, we construct a general Hamiltonian describing both bulk and surface states and use its low-energy limit to obtain expressions for the surface states at the zero-energy Dirac point ($\bar{\Gamma}$ point in the 2D BZ). Due to the presence of spin and orbital degrees of freedom and the TRS of the Hamiltonian, the latter are degenerated at this special point of the BZ. This approach can be regarded as doing degenerate perturbation theory in terms of the in-plane momentum $\mathbf{k}_{\parallel}$, whereby the parallel perturbation Hamiltonian is then projected onto the basis of the degenerate ground states, yielding the effective Hamiltonian for the states localized near the surface. In the end, the full expression for the effective Hamiltonian, corrected for the energy of the unperturbed states, and its spectrum are given by:

$$H_{2D}^{S}(\mathbf{k}_{\parallel}) = \left(h_0^{zo}(\mathbf{k}_{\parallel}) + B_0 \frac{-h_5^{zo}(\mathbf{k}_{\parallel}) + h_5^{\Gamma}}{B_{11}}\right)\hat{I}_2 + \sqrt{1 - \frac{B_0^2}{B_{11}^2}}\left[h_1^{zo}(\mathbf{k}_{\parallel})\sigma_x + h_2^{zo}(\mathbf{k}_{\parallel})\sigma_y + h_3^{zo}(\mathbf{k}_{\parallel})\sigma_z\right], \tag{17}$$

where B_0, B_{11}, and other values of A_0 are the TBM parameters of Bi_2Se_3 reported in Ref. [32]. The eigen-energies or bands of the TSSs Hamiltonian (17) are:

$$\mathscr{E}_{2D}^{S;\pm}(\mathbf{k}_{\parallel}) = 6A_0 + h_0^{zo}(\mathbf{k}_{\parallel}) - h_0^{\Gamma} + \frac{B_0}{B_{11}}\left(6B_{11} - h_5^{zo} + h_5^{\Gamma}\right) \pm \frac{\sqrt{\left(-B_0^2 + B_{11}^2\right)\sum_{i=1}^{3}\left(h_i^{zo}(\mathbf{k}_{\parallel})\right)^2}}{B_{11}}, \tag{18}$$

where ($\pm$) correspond to the lower ($-$) or the upper ($+$) bands or Dirac codes, respectively. **Figure 6(b)** shows the TSSs for the valence and conduction bands; about the Γ point we observe a null band gap in comparison with the bulk. This clearly indicates the surface states host spin electron currents. The surface eigenstates defined over the entire 2D surface BZ read:

$$|\Psi_{S,}^{\pm}(\mathbf{k}_{\parallel})\rangle = \mathcal{N}_S^{\pm}(\mathbf{k}_{\parallel}) f(\mathbf{k}_{\parallel}) \begin{pmatrix} h_3^{zo}(\mathbf{k}_{\parallel}) \pm \sqrt{\sum_{i=1}^{3}\left(h_i^{zo}(\mathbf{k}_{\parallel})\right)^2} \\ h_1(\mathbf{k}_{\parallel}) + ih_2(\mathbf{k}_{\parallel}) \end{pmatrix}, \tag{19}$$

with $f(\mathbf{k}_{\parallel}) = \left(h_1^{zo} + ih_2^{zo}\right)^{-1}$ and a normalization "constant" factor $\mathcal{N}_S^{\pm}(\mathbf{k}_{\parallel})$. The spin polarization refers to the degree to which the electron spin are aligned in a particular direction. For the topological surface states, this spin polarization is defined as $\langle\sigma_i\rangle_{\pm} = \langle\Psi_S^{\pm}|\sigma_i|\Psi_S^{\pm}\rangle$ the expectation value of the Pauli matrix σ_i with $i = 1,2,3$. **Figure 6(c)** and **(d)** exhibits the topological spin polarization for the lower and upper Dirac cones of the TSSs: $\{\langle\sigma_x\rangle_{\pm}, \langle\sigma_y\rangle_{\pm}, \langle\sigma_z\rangle_{\pm}\}$. The vectorial field for the *spin in plane* polarization denotes a vortex on contra-clockwise direction for the lower band, and the oppose behavior (clockwise rotation) occurs for the upper band. The spin *out of plane* polarization is observed in colorful scale, being dominant "far" for the $\bar{\Gamma}$ point but small abound the highly symmetric $\bar{\Gamma}$ point.

2.3.3 High harmonic generation from 3D topological insulators

We have built the necessary tools to compute the main ingredients and evaluate the electronic currents for a 3D TI such as Bi_2Se_3. Similar to the previous Section 2.2.1, we compute the HHG spectra firstly by solving numerically the time-dependent density matrix and secondly computing the Fourier components of the charge currents for both the topological bulk and surface states.

In **Figure 7**, we show the HHG spectra for the TBSs (a) and TSSs (b), driven by a right-handed circularly polarized laser. For bulk states, the harmonic emission shows clearly the sixfold symmetry and its corresponding selection rule, which allows harmonic orders of $(6n \pm 1)$ for co-rotating (magenta color) and contra-rotating (light blue). The highest harmonic emitted by the medium is till the 7th order and follows the same rotation that the driven laser: right-handed circularly polarized.

On another hand, the TSS emission exhibits a threefold symmetry and its expected selection rule satisfies $(3n \pm 1)$. Additionally, an interesting *large harmonic intensity emission* for the TSSs in comparison with the emission from TBSs is observed, which suggests this tendency (the harmonic emission from the topological surface states) of the HHG spectrum can encode structural information of the topological surface states. We can ask whether or not this signature is good enough to distinguish topological surface state. To answer this question, we perform theoretical calculations of the HHG signal as a function of the ellipticity of the driven laser.

The theoretical calculations are shown in **Figure 8** for both the topological bulk states and topological surface states as a function of the harmonic orders (HO): HO5, HO7, and HO9. The ellipticity dependence for the bulk states exhibits in the selected harmonic orders follows a trivial behavior: The maximum of the *intensity yield* occurs for linearly polarized driving lasers but it is minima for circularly polarized lasers. However and on the contrary, the topological surface state and its emission for HO5 and HO7 show totally oppose behavior: Their intensity emission is *large* for circularly

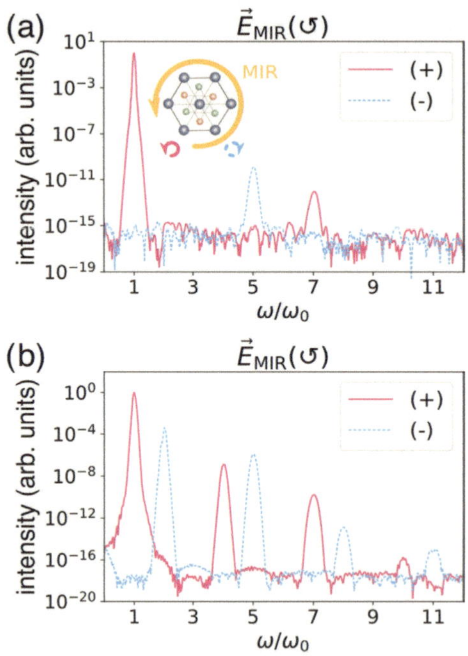

Figure 7.
High-order harmonic spectra from bismuth selenide: Bi_2Se_3. High photon energy emission driven by circularly polarized fields, for both bulk (a) and surface states (b). The relative orientation of the (111) surface (real space) and the MIR polarization vector is represented in the inset, whereby the propagation direction points toward the reader. The MIR pulse is left-circularly polarized with I_o = 0.0025 TW/cm^2 and a FWHM duration of 12 cycles. The helicity of the emitted harmonics is encoded in color: Left, co-rotating orders are shown in magenta (full lines), right, counter-rotating in cyan (dashed lines). Bulk and surface states obey different selection rules: $\omega_n = (6n \pm 1)\omega_o$ for the bulk (a) versus $\omega_n = (3n \pm 1)\omega_o$ (ω_o is the central frequency of the driving laser) for the surface (b). In both calculations, the dephasing time is set at T_2 = 1.25 fs.

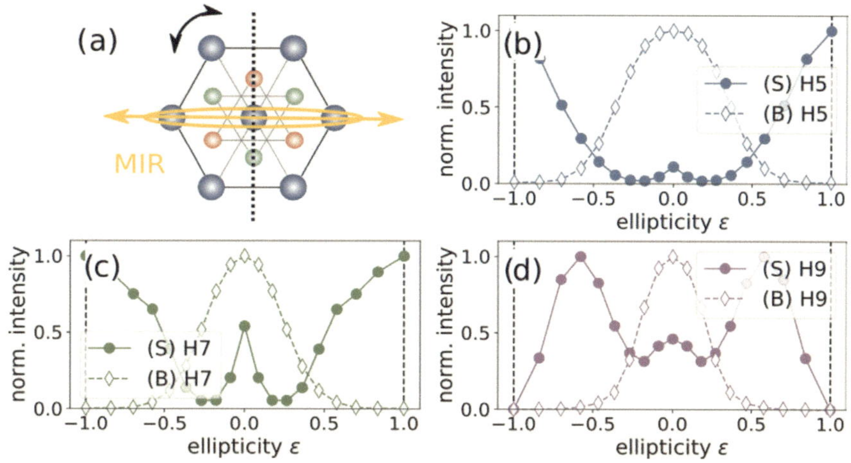

Figure 8.
Ellipticity dependence of the high-order harmonic emission from Bi_2Se_3. (a) Sketch of the excitation geometry, where the major axis of the MIR ellipse (orange) remains perpendicular to the mirror axis (dashed line) of the crystal through the measurement. (b)–(d) show the calculated ellipticity response for harmonic orders 5, 7, and 9. The calculations (12-cycle Gaussian pulse, I_0 = 0.0085 TW/cm², λ_{MIR} = 7.5 µm, T_2 = 1.25 fs) pertain to the contributions from the bulk ((B), dashed lines, diamonds) and from the surface states ((S), solid lines, circles).

polarized in comparison with the intensity yield produced by linearly polarized driving lasers.

This observation clearly demonstrates an anomalous feature of the intensity dependent of the harmonic emission as a function of the ellipticity. We re-call this distinguishes observation as *anomalous ellipticity dependence* of the high harmonic generation from the topological surface states of Bi_2Se_3.

2.3.4 Experimental observation of the topological anomalous ellipticity dependence

The high harmonic generation process from 3D topological insulators was also experimentally reported in Refs. [35, 36]. They produced and attributed the harmonic generation to the topological surface states and its spin-lock currents. Additionally, fractional harmonics were observed by Schmid et al. However, those experimental observation did not describe a clear signature of topology in the HHG emission. In this section, we show the main results of the experimental realization for the characterization of the topological surface states in Bi_2Se_3 reported in Ref. [33].

In **Figure 9** the experimental measurements of the ellipticity dependence of HO7, HO8, and HO9 are depicted: the ellipticity for the linearly polarized (LP) laser $\varepsilon = 0$, while for right-handed circularly polarized laser (RCP) is $\varepsilon = +1$ and for left-handed circularly polarized laser (LCP) $\varepsilon = -1$.

The HO7 shows an interesting anomalous behavior theoretically predicted above: The intensity yield is larger for CPL that LP laser drivers. However, for the HO8 and HO9 the harmonic yield varies across the ellipticity axis and trends to zero for $\varepsilon = \pm 1$. The theoretical calculations are shown in the second arrow of **Figure 9 (d–f)**. We can appreciate that for the special harmonic 7th the surface emission reproduces qualitatively the experimental result. In contrast, the bulk emission does not reproduce the main tendency of the experimental results. This impressively suggests that the anomalous ellipticity of the HO7 can encode information of the topological surface states

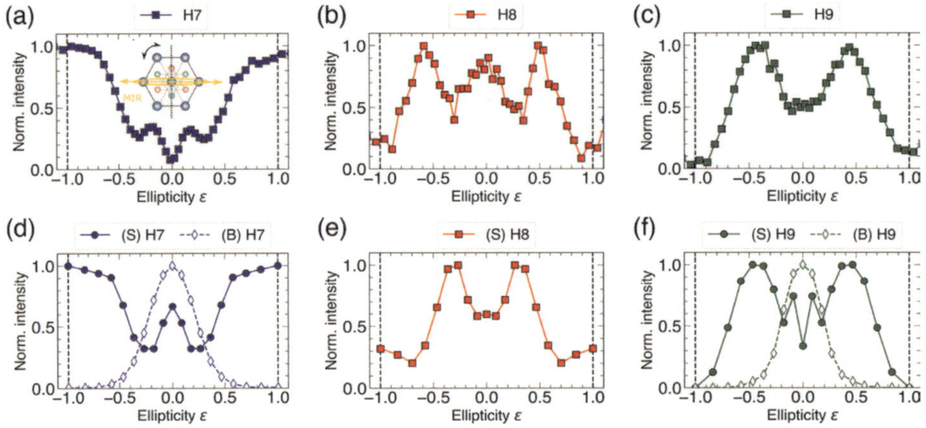

Figure 9.
Ellipticity dependence of the high-order harmonic emission: experimental observation. (a–c) Experimental measurement of the ellipticity-dependent response of HHG spectra at normal incident. The inset shows the experimental geometry, where the major axis of the MIR ellipse (indicated by the orange arrow) is kept orthogonal to the mirror axis (dashed black line) of the crystal throughout the measurement. (d–f) Calculated ellipticity response (for a 12-cycle Gaussian pulse, I_o = 0.0075 TW/cm^2 at ε = 0). The contributions of the bulk ("B," dashed lines, empty symbols) and the surface ("S," solid lines, filled symbols) are shown separately. Crystal symmetry prohibits HO9 for circular polarization.

similar to what it is done by Angular-Resolved PhotoElectron Spectroscopy (ARPES) measurements.

The mechanism behind this process is related to the laser-topological dipole matrix element coupling and the vectorial vortex of this dipolar moment in the BZ [33]. The electron is excited *via* tunneling from the valence band to the conduction band by the dipolar coupling, and both the intraband and the interband currents are driven by the strong electric field in the BZ; at the emission time, the electron comes back from the valence band converting the accumulating energy in a high photon. This photon contains information of the topological surface states by the dipole-moment coupling in the emission process. This justifies the fact that the anomalous ellipticity can map the topology of Bi_2Se_3.

This experimental observation lays the foundations to propuse the anomalous ellipticity to map topological phase transitions in 3D topological insulators as reported in Ref. [37]. In **Figure 10**, we depict the energy dispersion for TSSs measured *via* ARPES technique for pure Bi_2Se_3 and doped with In in 20 %$(Bi_{0.8}In_{0.2})Se_3$. We observe the topologically nontrivial Dirac cone in case of pure phase of Bi_2Se_3—the electrons can cross from the valence band to the conduction band; then, the crystal conducts carriers at the surface. However, once the material is doped in a 20 % the topologically nontrivial phase is lost—this clearly indicates that a topological phase transition takes place under the magnetic doping In element.

On another hand and in order to prove the topological phase transitions can be mapped by the high harmonic process, the HHG ellipticity dependence exhibits a clear enhancement of the harmonic intensity yield for circularly polarized laser. On the contrary, for a linear polarized driven laser the harmonic yield is minimum. Interestingly, once the Bi_2Se_3 is doped by In, the ellipticity dependence of the harmonic yield shows the standard behavior of a gaussian-like structure which is usually experimentally observed in MoS_2. This clearly demonstrates that the anomalous ellipticity measurement can map topological phase transitions. Therefore, one can argue that HHG is

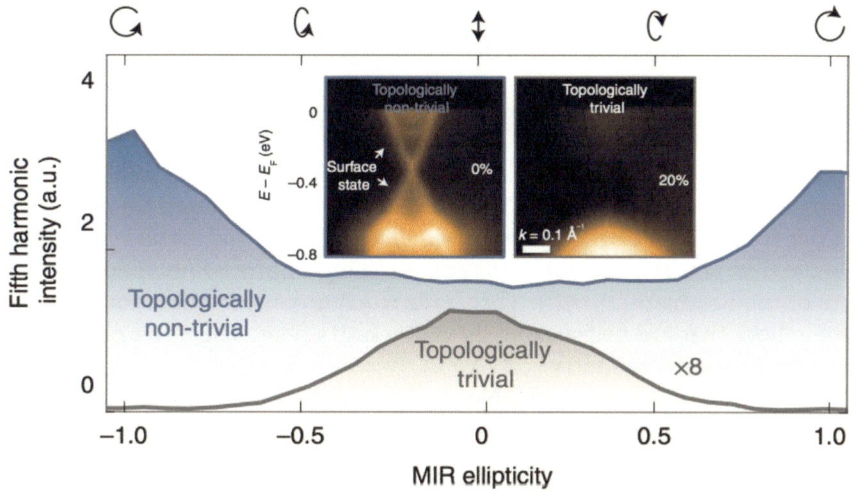

Figure 10.
Experimental observation of proving topological phase transition using HHG. Inset: Measured band structure using ARPES, showing the topologically protected surface state for the pure Bi_2Se_3 sample (left), which vanishes for high indium concentration (right). The intensity of the HO5 as a function of the MIR ellipticity ε is measured for both phases. A strong high harmonic enhancement is observed for the topologically non-trivial sample, especially for circular excitation. Note that intensity of the topologically trivial sample has been up-scaled by a factor of 8.

an capable observable to track topological phases and transitions in 3D topological insulator by measuring the ellipticity dependence of the harmonic orders. We have experimental evidences that suggest HHG thus can be used to diagnose topological materials. Additionally, the process of producing high harmonic from crystals does not damage the material. This is important for future technological application in the process of testing and validating topological materials.

2.4 Induced lightwave topological phases in hexagonal boron nitride

So far, we have discussed the profound implications that the HHG has to feature topological phases and transitions in topological materials and its remarkable experimental realization. These impressive theoretical and experimental efforts are important in promoting HHG as a new technological avenue to diagnose the topologically nontrivial phase of TIs. An additional question one can wonder is whether or not the strong field of ultrafast lasers can be used to control the topological phases and transitions optically. In order words, we propose the hypothesis that the lightwave can manipulate or engineer the electronic features of the material in such a way it can cross a topological phase transition.

In that framework, Jiménez et al. theoretically proposed the possibility to engineering the complex next-nearest neighbor (NNN) parameter by using bicircular laser electric field—a kind of Floquet topological insulators [38]. Floquet theory predicts the possibility to manipulate the complex NNN hopping parameters by employing circularly polarized laser-driven fields. Here, we discuss the application of this idea in the context of ultrafast laser pulses to induce a topological phase transition well controlled at the sub-cycle timescale by tailoring the laser-field characteristics. Actually, the study of Floquet topological states also has been carried out in "graphene" and used the high harmonic circular-dichroism to diagnose this state [39].

A very recent experimental observation performed by Mitra et al. [40] demonstrated that the lightwave can manipulate the electronic structure of a monolayer hexagonal lattice of boron nitride (hBN). By a two-color experimental scheme with a counter-rotating bicircular laser-light (see Eq. (20)) one can engineer the complex NNN hopping parameter of hBN then twisting the light waveform with respect to the lattice results in optical control of *time-reversal symmetry* breaking and the realization of the topological Haldane model. The parameters of the effective Haldane Hamiltonian are controlled by rotating the light wavefront. This enables ultrafast switching between band structure configurations and allowing unprecedented control over the structure of the bandgap [38, 40]. This leads to asymmetric population between quantum valleys that are measurable by valley Hall effect. The latter can be detected by optical harmonic polarimetry.

The counter-rotating bicircular field, widely used for controlling strong-field processes in atoms is defined as:

$$\mathbf{E}_{\mathrm{L}}(t) = \hat{\mathbf{e}}_x[E_1\cos(\omega_0 t) - E_2\cos(2\omega_0 t + \varphi_0)] + \hat{\mathbf{e}}_y[E_1\cos(\omega_0 t) + E_2\cos(2\omega_0 t + \varphi_0)], \tag{20}$$

where E_1 and E_2 are the field strengths of the fundamental ω_0 and second harmonic $2\omega_0$, respectively, and φ_0 is the sub-cycle phase delay between the two drivers that controls the orientation of the pulse in space. This is the key to engineering the complex NNN hopping parameter of Haldane model or breaks $\mathcal{T}$ symmetry.

The experiment measures the helicity-resolved third-harmonic $3\omega_0$ yield of a time-delayed linearly polarized probe as a function of pump trefoil rotation φ_0. It was performed by photodiode current measurements during the illumination of hBN sample. Thus, Mitra et al. show the observation of Haldane topological model controlling the optical shape of the bicircular pulses.

This work also demonstrate a new regime of valleytronics, namely the far-off resonant regime for the first time. Traditionally, for valleytronics one uses resonant circulary polarized light. Depending on the helicity of the light, the electron from the valence band edge gets excited to either K or K′ valleys, which is envisioned to be used for binary information processing. Mitra et al. showed similar end results by using far-off resonant light, utilizing the principle of time-reversal symmetry breaking.

3. Conclusions

We have introduced the concepts of topological phases of matter and their diagnosis and control by the manipulation of the (i) light emission produced *via* the laser-crystal interaction, and (ii) electronic structure *via* pump-prope pulses. This new phase in condensed matter physics, topological materials, exhibits insulating bulk states and conducting edge (surface) states *protected* by symmetries, in particular for topological insulators, the time-reversal symmetry and the topological invariants.

We reviewed the latest progress in the field of Attosecond Physics applied to study, diagnosis, and control topological phases and transitions in different classes of topological materials, for instance: Chern insulators, 2D and 3D topological insulators, and the lightwave control of topological phase transition in hexagonal boron nitride.

We find shreds of evidence suggesting that the helicity, circular-dichroism, and anomalous ellipticity dependence produced from high-order harmonic generation driven from linearly to circularly polarized laser fields are useful quantities to prove qualitatively the topological phases and transitions of matter. Both theoretical and experimental observations are presented across the sections of this chapter. Finally, we show how the lightwave can drive the energetic band structure by manipulating second hopping parameters. The study of proving all-optical proof of topological insulators and controlling topological phases of matters is still an emerging field within the application of attosecond science to condensed matter physics. We believe there are still open questions such as how the electron dynamics at edge states in the Quantum Hall Spin Effect or TIs is recorded at its natural timescale, which can be addressed using attosecond science tools. Since attosecond pulse provides sub-femtosecond timescale resolution, it is the most natural or logical way to face this question. The other important open question is how the topological invariant is clearly obtained from the experimental results. Still, our theoretical prediction of HHG from topological materials has not retrieved the topological invariant one-to-one by a closed formula like the QHEs.

Acknowledgements

A.C. acknowledges financial support form the Sistema Nacional de Investigación (SNI) and the Secretaría Nacional de Ciencia, Innovación y Tecnología (SENACYT) of Panama. A.C. thanks Dr. Shubhadeep Biswas SLAC National Accelerator Laboratory, Stanford University, for fructiferous discussions.

Conflict of interest

The author declares no conflict of interest.

Appendix

In the following, we define the auxiliary functions $h_i(\mathbf{k})$ used in the TBM Hamiltonian defined in eq. (13)

$$h_0(\mathbf{k}) = 2A_0 \sum_{i=1}^{3} \cos(\mathbf{k}\cdot\mathbf{a}_i) + 2B_0 \sum_{i=1}^{3} \cos(\mathbf{k}\cdot\mathbf{b}_i) \tag{21}$$

$$\begin{aligned} h_1(\mathbf{k}) &= -2A_{14}\sin[\omega(\sin(\mathbf{k}\cdot\mathbf{a}_2) - \sin(\mathbf{k}\cdot\mathbf{a}_3))] \\ &+2B_{14}[\sin(\mathbf{k}\cdot\mathbf{b}_1) + \cos(\omega(\sin(\mathbf{k}\cdot\mathbf{b}_2) + \sin(\mathbf{k}\cdot\mathbf{b}_3))], \\ h_2(\mathbf{k}) &= -2B_{14}[\sin(\omega(\sin(\mathbf{k}\cdot\mathbf{b}_2,)) - \sin(\mathbf{k}\cdot\mathbf{b}_3,] \\ &-2A_{14}[\sin(\mathbf{k}\cdot\mathbf{a}_1) + \cos(\omega(\sin(\mathbf{k}\cdot\mathbf{a}_2) + \sin(\mathbf{k}\cdot\mathbf{a}_3)))], \end{aligned} \tag{22}$$

$$h_3(\mathbf{k}) = 2A_{12} \sum_{i=1}^{3} \sin(\mathbf{k}\cdot\mathbf{a}_i) \tag{23}$$

$$h_4(\mathbf{k}) = -2B_{12}\sum_{i=1}^{3}\sin(\mathbf{k}\cdot\mathbf{b}_i) \tag{24}$$

$$h_5(\mathbf{k}) = 2A_{11}\sum_{i=1}^{3}\sin(\mathbf{k}\cdot\mathbf{a}_i) + B_{11}\sum_{i=1}^{3}\sin(\mathbf{k}\cdot\mathbf{b}_i) + m_{11}, \tag{25}$$

where $\omega = -\frac{2\pi}{3}$. m_{11} in $h_5(\mathbf{k})$ controls the band inversion mechanism behind TIs.

Author details

Alexis Chacón[1,2,3]

1 Departamento de Física, Área de Física, Universidad de Panamá, Ciudad Universitaria, Panama, Panama

2 Sistema Nacional de Investigación, Ciudad del Saber, Clayton Panama, Panama

3 Parque Científico y Tecnológico, Universidad Autónoma de Chiriquí, Ciudad Universitaria, David, Panama

*Address all correspondence to: alexis.chacon-s@up.ac.pa

References

[1] Tokura Y. Quantum materials at the crossroads of strong correlation and topology. Nature Materials. 2022;**21**(9): 971-973. DOI: 10.1038/s41563-022-01339-6 [Accessed: 24 August 2022]

[2] Klitzing KV, Dorda G, Pepper M. New method for high-accuracy determination of the fine-structure constant based on quantized hall resistance. Physical Review Letters. 1980;**45**(6):494-497. DOI: 10.1103/PhysRevLett.45.494 [Accessed: May 30, 1980]

[3] Stormer H. Nobel lecture: The fractional quantum hall effect. Reviews of Modern Physics. 1999;**71**(4):875-889. DOI: 10.1103/RevModPhys.71.875 [Accessed: 1 July 1999]

[4] Thouless DJ, Kohmoto M, Nightingale MP, den Nijs M. Quantized hall conductance in a two-dimensional periodic potential. Physical Review Letters. 1982;**49**(6):405-408. DOI: 10.1103/PhysRevLett.49.405 [Accessed: 9 August 1982]

[5] Hasan MZ, Kane CL. Colloquium: Topological insulators. Review of Modern Physics. 2010;**82**(4):3045-3067. DOI: 10.1103/RevModPhys.82.3045 [Accessed: 8 November 2010]

[6] Haldane FDM. Model for a quantum hall effect without Landau levels: Condensed-matter realization of the "parity anomaly". Physical Review Letters. 1988;**61**(18):2015-2018. DOI: 10.1103/PhysRevLett.61.2015 [Accessed: 31 October 1988]

[7] Xiao D, Chang M-C, Niu Q. Berry phase effects on electronic properties. Review of Modern Physics. 2010;**82**(3): 1959-2007. DOI: 10.1103/RevModPhys.82.1959 [Accessed: 6 July 2010]

[8] Jotzu G, Messer M, Desbuquois R, et al. Experimental realization of the topological Haldane model with ultracold fermions. Nature. 2014; **515**(7526):237-240. DOI: 10.1038/nature13915 [Accessed: 12 November 2014]

[9] Zhao W, Kang K, Zhang Y, et al. Realization of the Haldane Chern insulator in a moiré lattice. Nature Physics. 2024;**20**(2):275-280. DOI: 10.1038/s41567-023-02284-0 [Accessed: 05 January 2024]

[10] Kane CL, Mele EJ. Quantum spin hall effect in graphene. Physical Review Letters. 2005;**95**(22):226801-1-226801-4. DOI: 10.1103/PhysRevLett.95.226801 [Accessed: 23 November 2005]

[11] König M, Wiedmann S, Brüne C, et al. Quantum spin hall insulator state in HgTe quantum wells. Science. 2007; **318**(5851):766-770. Available from: https://www.science.org/doi/10.1126/science.1148047 [Accessed: 02 November 2007]

[12] Shumiya N, Hossain MS, Yin JX, et al. Evidence of a room-temperature quantum spin hall edge state in a higher-order topological insulator. Nature Mater. 2022;**21**(10):1111-1115. DOI: 10.1038/s41563-022-01304-3 [Accessed: 14 July 2022]

[13] Kane CL, Mele EJ. Z_2 topological order and the quantum spin hall effect. Physical Review Letters. 2005;**95**(14): 146802-1-146802-4. DOI: 10.1103/PhysRevLett.95.146802 [Accessed: 28 September 2005]

[14] Bernevig BA, Hughes TL, Zhang SC. Quantum spin hall effect and topological phase in HgTe quantum well. Science.

2006;**314**(5806):1757-1761. DOI: 10.1126/science.1133734 [Accessed: 14 November 2016]

[15] Zhang H, Liu CX, Qi XL, et al. Topological insulators in Bi_2Se_3, Bi_2Te_3 and Sb_2Te_3 with a single Dirac cone on the surface. Nature Physics. 2009;**5**(6): 438-442. DOI: 10.1038/nphys1270 [Accessed: 10 May 2009]

[16] Krausz F, Ivanov M. Attosecond physics. Review of Modern Physics. 2009;**81**(1):163-234. DOI: 10.1103/RevModPhys.81.163 [Accessed: 2 February 2009]

[17] Hentschel M, Kienberger R, Spielmann C, et al. Attosecond metrology. Nature. 2001;**6863**(414): 509-513. DOI: 10.1038/35107000 [Accessed: 29 November 2001]

[18] Ferray M, L'Huillier A, Li XF, Lompre LA, Mainfray G, Manus C. Multiple-harmonic conversion of 1064 nm radiation in rare gases. Journal of Physics B: Atomic, Molecular and Optical Physics. 1988;**3**(21):L31-L35. DOI: 10.1088/0953-4075/21/3/001

[19] Li XF, L'Huillier A, Ferray M, Lompré LA, Mainfray G. Multiple-harmonic generation in rare gases at high laser intensity. Physical Review A. 1989; **39**(11):5751-5761. DOI: 10.1103/PhysRevA.39.5751 [Accessed: 3 January 1989]

[20] Corkum PB. Plasma perspective on strong field multiphoton ionization. Physical Review Letters. 1993;**13**(71): 1994-1997. DOI: 10.1103/PhysRevLett.71.1994 [Accessed: 9 February 1993]

[21] Lewenstein M, Ph Balcou MY, Ivanov AL'H, Corkum PB. Theory of high-harmonic generation by low-frequency laser fields. Physical Review A. 1994;**3**(49):2117-2132. DOI: 10.1103/PhysRevA.49.2117 [Accessed: 19 August 1993]

[22] Itatani J, Levesque J, Zeidler D, Niikura H, Pépin H, Kieffer JC, et al. Tomographic imaging of molecular orbitals. Nature. 2004;**7019**(432): 867-871. DOI: 10.1038/nature03183 [Accessed: 16 December 2004]

[23] Ghimire S, DiChiara DA, Sistrunk E, Agostini P, DiMauro LF, Reis DA. Observation of high-order harmonic generation in a bulk crystal. Nature Physics. 2011;**2**(7):138-141. DOI: 10.1038/nature03183 [Accessed: 5 December 2010]

[24] Luu TT, Garg M, Yu Kruchinin S, Moulet A, Th Hassan M, Goulielmakis E. Extreme ultraviolet high-harmonic spectroscopy of solids. Nature. 2015; **7553**(521):498-502. DOI: 10.1038/nature14456 [Accessed: 27 May 2015]

[25] Vampa G, Hammond TJ, Thiré N, Schmidt BE, Légaré F, McDonald CR, et al. Linking high harmonics from gases and solids. Nature. 2015;**7557**(522): 462-464. DOI: 10.1038/nature14517 [Accessed: 24 June 2015]

[26] Vampa G, McDonald CR, Orlando G, Klug DD, Corkum PB, Brabec T. Theoretical analysis of high harmonic generation in solids. Physical Review Letters. 2014;**7**(113):073901–1-073901–5. DOI: 10.1103/PhysRevLett.113.073901 [Accessed: 13 August 2014]

[27] Schubert O, Hohenleutner M, Langer F, Urbanek B, Lange C, Huttner U, et al. Sub-cycle control of terahertz high-harmonic generation by dynamical Bloch oscsillations. Nature Photonics. 2014;**2**(8):119-123. DOI: 10.1038/nphoton.2013.349 [Accessed: 19 January 2014]

[28] Liu H, Li Y, You YS, Ghimire S, Heinz TF, Reis DA. High-harmonic generation from an atomically thin semiconductor. Nature Physics. 2017; **3**(13):262-265. DOI: 10.1038/nphoton.2013.349 [Accessed: 14 November 2016]

[29] Bauer D, Hansen KK. High-harmonic generation in solids with and without topological edge states. Physical Review Letters. 2018;**17**(120):177401-1-177401-6. DOI: 10.1103/PhysRevLett.120.177401 [Accessed: 24 April 2018]

[30] Silva REF, Jiménez-Galán A, Amorim B, Smirnova O, Ivanov M. Topological strong-field physics on sub-laser-cycle timescale. Nature Photonics. 2019;**12**(13):849-854. DOI: 10.1038/s41566-019-0516-1 [Accessed: 23 September 2019]

[31] Chacón A, Kim D, Zhu W, Kelly SP, Dauphin A, Pisanty E, et al. Circular dichroism in higher-order harmonic generation: Heralding topological phases and transitions in Chern insulators. Physical Review B. 2020;**13**(102): 134115-1-134115-13411520. DOI: 10.1103/PhysRevB.102.134115 [Accessed: 21 October 2020]

[32] Baykusheva D, Chacón A, Kim D, Kim DE, Reis DA, Ghimire S. Strong-field physics in three-dimensional topological insulators. Physical Review A. 2021;**2**(103):023101-1-023101-02310117. DOI: 10.1103/PhysRevA.103.023101 [Accessed: 2 February 2021]

[33] Baykusheva D, Chacón A, Lu J, Bailey TP, Sobota JA, Soifer H, et al. All-optical probe of three-dimensional topological insulators based on high-harmonic generation by circularly polarized laser fields. ACS Nano Letters. 2021;**21**(21):8970-8978. DOI: 10.1021/acs.nanolett.1c02145 [Accessed: 22 October 2021]

[34] Liu CX, Qi XL, Zhang HJ, Dai X, Fang Z, Zhang SC. Model Hamiltonian for topological insulators. Physical Review B. 2010;**4**(82):045122-1-19. DOI: 10.1103/PhysRevB.82.045122 [Accessed: 26 July 2010]

[35] Bai Y, Fei F, Wang S, Li N, Li X, Song F, et al. High-harmonic generation from topological surface states. Nature Physics. 2021;**3**(17):311-315. DOI: 10.1038/s41567-020-01052-8 [Accessed: 23 November 2020]

[36] Schmid CP, Weigl L, Grössing P, Junk V, Gorini C, Schlauderer S, et al. Tunable non-integer high-harmonic generation in a topological insulator. Nature. 2021;**7859**(593):385-390. DOI: 10.1038/s41586-021-03466-7 [Accessed: 19 May 2021]

[37] Heide C, Kobayashi Y, Baykusheva DR, Jain D, Sobota JA, Hashimoto M, et al. Probing topological phase transitions using high-harmonic generation. Nature Photonics. 2022; **9**(16):620-624. DOI: 10.1038/s41566-022-01050-7 [Accessed: 18 August 2022]

[38] Jiménez-Galán A, Silva REF, Smirnova O, Ivanov M. Lightwave control of topological properties in 2D materials for sub-cycle and non-resonant valley manipulation. Nature Photonics. 2020). (7/8): 2 Available from: http://manfredfehr.com.br/qualidade1.pdf;**14**: 728-732. DOI: 10.1038/s41566-020-00717-3 [Accessed: December 06, 2016]

[39] Kim J, Kim D, Kim DE, Chacón A. Circular dichroism in Floquet Chern insulator via high-order harmonics spectroscopy. Journal of Physics: Condensed Matter. 2023;**36**(3):035701.

DOI: 10.1088/1361-648X/ad0015 [Accessed: 20 October 2023]

[40] Mitra S, Jiménez-Galán A, Aulich M, Neuhaus M, Silva REF, Pervak V, et al. Light-wave-controlled Haldane model in monolayer hexagonal boron nitride. Nature. 2024;**628**:752. DOI: 10.1038/s41586-024-07244-z [Accessed: 15 April, 2024]

Chapter 4

Topology and the Quantum Hall Effects

Paul Bracken

Abstract

The quantum Hall effects are an excellent example of physical systems where topology plays a major role in accounting for the physical observations. It is shown that the conductivity that appears in the quantum Hall effect is a topological invariant. It is illustrated how a fiber bundle over a torus can be constructed producing a geometry in which the system can be referred. The fractional effect can be studied by introducing homotopy and associated braid groups. Filling fractions can be obtained as a consequence of commensurability relations.

Keywords: topology, gauge, group, manifold, connection, bundle, fiber, path integral

1. Introduction

It has been observed experimentally that electrons restricted to move in a two-dimensional plane in a strong magnetic field give rise to many phenomena known collectively as the quantum Hall effect. The Hall conductivity takes quantized values

$$\sigma_{xy} = \frac{e^2}{2\pi\hbar}\nu. \tag{1}$$

At first it was found that ν to extraordinary precision takes integer values. Subsequently it was discovered that ν is not restricted to integers but may also take specific rational values, such as $1/3$ and $2/5$, as well as many different fractions that have been observed. It is the interactions between electrons that result in a highly correlated quantum state or a new state of matter. To explain this new phase, something novel is required, and topology has a role to play in understanding many body systems. The study of physical phenomena in which topology plays a significant role in their understanding is a subject of great current interest. Topology is concerned with properties of an object or system that are not changed by continuous distortions. Many areas of solid-state physics have been influenced by topology, but the one in particular investigated here is the quantum Hall effects [1–3].

Topology has assumed increasing importance and is regarded as fundamental in accounting for many phenomena in modern physics, in particular, global topological effects. Methods of a largely topological character are also applied in areas such as relativistic physics, field theory, and even quantum gravity [4]. Some topological effects are evident in path integral quantization but are not explicit in the Schrödinger equation without additional restrictions. Path integral quantization is explicitly nonlocal in distinction to Schrödinger quantization, which is a differential equation. The study of topological defects was developed within field theory and has led to the concept of objects with a topological charge, such as instantons. Topological methods have been extremely important in a very basic aspect: the quantization of systems containing many identical particles. The experimental discovery of the quantum Hall effect has contributed to the recognition of the role of the spatial topology of the manifold, which describes the system as far as determining the characteristics of the quantum particles undergoing dynamics there [5–10].

The integer (IQHE) and fractional quantum Hall effects (FQHE) result as a consequence of the cyclotron commensurability for 2D charged, interacting particles under the presence of a magnetic field. IQHE is not equivalent to the complete filling of Landau levels (LL), in particular, the lowest LL. IQHE is a strongly correlated state of interacting particles in 2D at integer filling of the Landau levels. In the FQHE, it must be distinguished between gaps in Landau level structure, a single particle effect, and gaps or activation energies related to multi-particle correlations of interacting electrons. The latter form of both effects is conditioned by the electron interaction, and the former is a single particle effect.

The first hint with regard to the quantization of the Hall conductance was motivated by a widely known but unpublished work by Block in the thirties [11–13]. It states the free energy of the equilibrium state for a loop or other non-simply connected piece of conductor is a periodic function of the flux Φ enclosed by the loop with period h/e, so the current is periodic and has zero average. Laughlin's argument which is based on a generalization of Bohm's version of Block's theorem. It leads to the conclusion that the Hall conductance is quantized.

Usually one wants to remove disorder to see the underlying physics. The longitudinal resistivity also gives a surprise. When the off-diagonal element sits on a plateaux, the longitudinal resistivity vanishes. It spikes when the off-diagonal jumps to the next plateaux. However, plateaux emerge at fractional values. This is the fractional quantum Hall effect. On the plateaux, the Hall resistivity takes a simple form implied by (1), and ν is a rational number $\nu \in \mathbb{Q}$, and not all fractions appear. It could be said that the reason for plateaus that are observed in the Hall conductivity is the effect of the re-entrance of the correlated state.

One particular theoretical description that makes the similarity between the integer and fractional quantum Hall effects explicit is the composite fermion (CF) theory of the FQHE [10]. It relates states of the interacting electron system at filling factor $\nu = \nu'/(p\nu' \pm 1)$ to states of noninteracting electrons at filling ν' by attaching an even number of p flux quanta to each electron. The magical filling factors of the interacting electron system are interpreted as filled Landau levels, $\nu' = k$, k an integer, of the CFs. Based on this approach, Jain was able to argue that transitions between two fractional quantum Hall plateaus fall into the universality class of the IQHE if these correspond to successive filled Landau levels of the CFs [11–15]. Treating a uniform density FQHE state as the vacuum, it appears as though an added electron fractionalizes into several fractionally charged *CF*-quasi-particles. Not just the charge of the electron fractionalize, but it happens as well to the statistics [14–21].

The objective is to discuss topological aspects of two-dimensional periodic systems in a magnetic field. In general the electronic properties of a topological material can be determined only by examining the complete set of states in an electronic band. That non-locality lends tremendous potential to topological materials. If a property is not defined locally, then it cannot be destabilized by local defects or fluctuations. It will be shown that the Hall conductance can be represented by means of a topological invariant, which is in fact an integer. The linear response or Nakano-Kubo formula for the Hall conductance is introduced. It is shown that a contribution of a single filled band to the conductivity for the integer effect is given by an integer associated with a topological invariant, a Chern number for a certain bundle. This physical situation can be described mathematically in terms of fiber bundles. The Dirac magnetic monopole bundle is a principal $U(1)$ bundle over a sphere S^2. This is very useful because it happens that topological invariants can be associated with these mathematical structures. This is done for the IQHE and it is outlined how this approach can be extended to the FQHE [22–24].

Some new results that do not seem to appear in CF theory are obtained by next introducing some ideas from algebraic topology. The main structures that arise from algebraic topology The rigorous inclusion of the indistinguishability of identical particles must be done in classical topological terms within the definition of multi-particle configuration space. For N indistinguishable particles it has the form $J_N = P_N/S_n$ where P_N is configuration space of an N-particle system. This is a quotient structure by S_N the permutation group of N-elements. Points in J_N are unified if they differ by numbering of particles only; in P_N space such points are different. The space J_N does not have an intuitive geometrical visualization unlike P_N.

Closed trajectory loops in J_N form disjoint classes of closed continuous trajectories starting from a certain point in J_N and finishing in the same point, closed multistrand loops in J_N joining the same positions of N particles distinctly numbered are called braids. These braids are open in P_N, but are closed loops in J_N. Closed loops in any topological space, arc-connected, $\mathcal{A}$ can be characterized in a topological sense by the first homotopy group $\pi_1(\mathcal{A})$ of the space. This group collects disjoint classes of inequivalent loops, which can be deformed one into the another cutting within the class but cannot be deformed between different classes. The space $\mathcal{A}$ is multiply connected if $\pi_1(\mathcal{A})$ is nontrivial. When the group is trivial $\pi_1(\mathcal{A}) = \{e\}$, space $\mathcal{A}$ is simply connected. When $\mathcal{A}$ is J_N, then $\pi_1(J_N)$ is called the braid group, and is usually nontrivial for $N \geq 2$. For a three-dimensional manifold M or higher, the braid group $\pi_1(J_N) = S_N$, the ordinary permutation group with $N!$ elements. For two-dimensional manifolds, the related braid groups are infinite and more complicated. For the $M = R^2$ the plane, the full braid group is the Artin group [25–30].

In other words, the π_1 homotopy group is the group structure of all classes of topologically nonequivalent trajectories that cannot be continuously transformed into one another by means of deformation. In this event, they are said to be nonhomotopic. Such groups are strongly dependent on the type of physical space in which the particles are located. The recovery of the Laughlin correlations in charged multi-particle $2D$ systems in a strong perpendicular magnetic field can be seen by means of cyclotron braid groups. It is demonstrated overall that the topological dissimilarity of planar systems is expressed in the form of new exotic physics [14, 15].

2. Quantum Hall effects

Let us begin by looking at a classical approach to the subject. The conductivity of a material is a measure of its response to an applied electric field, which if sufficiently small, relates the current density to the electric field, and defines conductivity σ

$$\mathbf{J} = \sigma \mathbf{E}. \tag{2}$$

The variable σ in (2) is called the conductivity. In the presence of a magnetic field, σ is a matrix known as the conductivity tensor

$$\sigma = \begin{pmatrix} \sigma_{xx} & \sigma_{xy} \\ -\sigma_{xy} & \sigma_{xx} \end{pmatrix} \tag{3}$$

The structure of σ follows from rotational invariance. The off-diagonal terms in the matrix are responsible for the Hall effect. When the electric field points along the x-axis, and the magnetic field B is aligned with the z axis, the resulting current will have components along both x and y directions. The fact that there exists a y component which depends on the Hall conductivity σ_H corresponds to the Hall effect. The components of $\mathbf{J}$ are then

$$j_x = \sigma E, \qquad j_y = \sigma_H E. \tag{4}$$

This can be thought of as a generalization of Ohm's law. Classical physics assumes the current is carried by a gas of noninteracting particles and if the charge carriers are electrons with charge $-e$ and electron number density n, σ_H is determined to be

$$\sigma_H = -\frac{nec}{B}. \tag{5}$$

When quantum mechanics is used to study electrons constrained to planar motion, the single particle energy is quantized. The energy spectrum is a regularly spaced sequence of Landau levels spaced by the amount $\hbar\omega_c$. The center of each of these plateaus occurs when the magnetic field takes the value

$$B = \frac{2\pi\hbar n}{\nu e} = \frac{n}{\nu}\Phi_0. \tag{6}$$

In (6), $\Phi_0 = 2\pi\hbar/e$ is known as the flux quantum. To compress the required density of states into the discrete spectrum, each level must have an enormous degeneracy. The degeneracy of each level is equal to the total flux through the plane $\Phi = hc/e$, and the Hall conductance is an integer multiple of e^2/h. Thus, if the Fermi energy lies in a gap between Landau levels, the Hall conductance is quantized in units of the conductance quantum e^2/h. It was subsequently discovered that under appropriate experimental conditions, conductance plateaux can also be observed as fractional multiples of e^2/h as well [19].

What is of interest is how the physics of two-dimensional Hall systems can be discussed topologically. To develop this further, it is fundamental to introduce mathematical concepts such as fiber bundles and Chern classes. Bundles very often have topological invariants associated with them. Every fiber bundle has associated base space and a fiber space. Electromagnetism is an example of a $U(1)$ bundle, so if one has a $U(1)$ bundle over the torus then $U(1)$ is the fiber and the torus is the base space. A copy of the fiber is attached at each point of the base space. In physical applications, the base space is the physical space, and the attached fiber is a space describing a degree of freedom or other property that a system or space has at that point. For a superconductor, the physical space is three-dimensional Euclidean space and order parameter a $U(1)$ phase.

A second important factor in constructing a bundle is the specification of a global geometrical way of assembling the fiber bundle in a smooth way. A bundle is obtained by attaching a line at each point on a circle, for example. The strip is the bundle, the circle is the base space at each point on it, and a perpendicular line is attached, the fiber. Another way to do this is to include a half-twist giving the Möbius bundle. The untwisted and twisted Möbius strips have the same base space and fiber, and hence, locally, they are identical. Bundles can be characterized by an integer, a topological invariant which is known as the Chern number [19]. Topological invariant means equivalent bundles have the same invariant Chern number. Mathematically, the Chern number is defined in terms of a connection. In this context, a connection specifies how to assign a reference phase to points along a curve against which other phase factors can be measured. The study of strongly correlated electron systems in two dimensions offers the possibility to look at global and topological properties that are not easily attained using other, more conventional methods. The long-distance physics of the ground state and its low-lying excitations is described by a new class of field theories called topological field theories in the integer Hall effect [8–10].

3. Electrons in a uniform magnetic field

In the study of electron bands, the Brillouin zone plays a kind of role as a geometric space. Closed surfaces in the zone can be thought of as geometric shapes that have an integer-valued index much like the genus of a surface. The Brillouin zone has effective periodic boundary conditions. Consider a unit cell or magnetic unit cell through which an integral multiple of magnetic flux passes. Bravais lattice vectors are taken, which assume the form

$$\mathbf{R} = nq\,\mathbf{a} + \mathbf{b}. \tag{7}$$

For any Bravais lattice vector $\mathbf{R}$, there is a translation operator $T_{\mathbf{R}}$ which, when operating on any smooth function $f(\mathbf{r})$ shifts $\mathbf{r}$ by a constant $\mathbf{R}$

$$T_{\mathbf{R}} f(\mathbf{r}) = f(\mathbf{r} + \mathbf{R}). \tag{8}$$

It has the explicit form,

$$T_{\mathbf{R}} = \exp\left[\frac{i}{\hbar}\mathbf{R}\cdot\mathbf{p}\right]. \tag{9}$$

Magnetic translations do not commute with each other in general since

$$T_{\mathbf{a}} \cdot T_{\mathbf{b}} = \exp(2\pi i \phi_{ab})\, T_{\mathbf{b}} \cdot T_{\mathbf{a}}, \tag{10}$$

where $\phi_{ab} = (eB/\hbar)ab$ is the number of magnetic flux lines in the unit cell. When ϕ_{ab} is a rational number, we have a subset of translations which commute with each other. Take an enlarged unit cell called a magnetic unit cell through which an integral magnetic flux passes so the Bravais lattice vectors are taken to have the form,

$$\mathbf{R}' = n(q\,\mathbf{a}) + m\,\mathbf{b}. \tag{11}$$

Then p magnetic flux quanta are in the magnetic unit cell, which is formed out of the vectors $q\,\mathbf{a}$ and $\mathbf{b}$. The magnetic translation operators $T_{\mathbf{R}'}$ which correspond to the new lattice vectors, commute with each other.

The Schrödinger equation for a two-dimensional noninteracting electron system in a uniform magnetic field perpendicular to the plane is

$$H\Psi = \left[\frac{1}{2m}(\mathbf{p} - e\mathbf{A})^2 + U(x,y)\right]\Psi = E\Psi. \tag{12}$$

The momentum $\mathbf{p} = -i\hbar\nabla$ and gauge potential $\mathbf{A}$ resides in the xy-plane. Let U be periodic so it satisfies $U(x+a,y) = U(x,y+b) = U(x,y)$.

The system is set to be invariant under a translation by a along x or by b along y-direction, but the Hamiltonian is not. This is due to the fact that the gauge potential $\mathbf{A}$ is not constant even though the magnetic field is uniform, so as gauge translation is needed to make it invariant. Define a translation operator T_R which operates as smooth functions by shifting the argument by $\mathbf{R}$, $T_R f(\mathbf{r}) = f(\mathbf{r}+\mathbf{R})$, or explicitly,

$$T_R = \exp\left(\frac{i}{\hbar}\mathbf{R}\cdot\mathbf{p}\right). \tag{13}$$

If T_R acts on (12), $U(\mathbf{r})$ is left invariant, but the gauge potential is transformed into $\mathbf{A}(\mathbf{r}+\mathbf{R})$. In general $\mathbf{A}(\mathbf{r})$ and $\mathbf{A}(\mathbf{r}+\mathbf{R})$ differ by a gradient of a scalar function in $\mathbf{B}$ is uniform

$$\mathbf{A}(\mathbf{r}) = \mathbf{A}(\mathbf{r}+\mathbf{R}) + \nabla g(\mathbf{r}). \tag{14}$$

Consider the magnetic translation operators

$$\hat{T}_R = \exp\left\{\frac{i}{\hbar}\mathbf{R}\cdot\left[\mathbf{p} + \frac{e}{2}(\mathbf{r}\times\mathbf{B})\right]\right\} = T_R\exp\left\{\frac{ie}{2\hbar}(\mathbf{B}\times\mathbf{R})\cdot\mathbf{r}\right\}. \tag{15}$$

In the symmetric gauge, $\hat{T}_R$ leaves the Hamiltonian invariant

$$\left[\hat{T}_R, H\right] = 0. \tag{16}$$

In this event, eigenstates that simultaneously diagonalize $\hat{T}_R$ and H are to be found. The magnetic translations do not commute with each other since

$$\hat{T}_a\hat{T}_b = \exp(2\pi i\phi)\hat{T}_b\hat{T}_a. \tag{17}$$

Here $\phi = (eB/h)ab$ is the number of magnetic flux lines in the magnetic cell. When ϕ is a rational number so that $\phi = p/q$, with p, q relatively prime integers, a subset of translations that commute with each other results.

An eigenfunction ψ which diagonalizes H and T_R simultaneously has the following eigenvalues under $\hat{T}_{qa}$ and $\hat{T}_b$

$$\hat{T}_{qa}\psi = e^{ik_1qa}\psi, \qquad \hat{T}_b\psi = e^{ik_2b}\psi, \tag{18}$$

where k_1 and k_2 are generalized momenta that can be restricted in the magnetic Brillouin zone $0 \le k_1 \le 2\pi/qa$, $0 \le k_2 \le 2\pi/b$. The eigenfunctions are given in terms of k_1 and k_2 in addition to the band index α in Block form

$$\psi_{k_1k_2}(x,y) = e^{ik_1x+ik_2y}\varphi^{\alpha}_{k_1k_2}(x,y). \tag{19}$$

On account of (18), function $\varphi^{\alpha}_{k_1k_2}$ has the translation property

$$\varphi^{\alpha}_{k_1k_2}(x+qa,y) = e^{-i\pi pq/b}\,\varphi^{\alpha}_{k_1k_2}(x,y), \qquad \varphi^{\alpha}_{k_1k_2}(x,y+b) = e^{i\pi px/qa}\,\varphi^{\alpha}_{k_1k_2}(x,y) \tag{20}$$

These are known as the generalized Block conditions. A gauge invariant, hence physically meaningful quantity, is the phase change around the boundary of the magnetic unit cell.

Another way to present this in gauge form is as follows. The operator form of a gauge transformation corresponding to Λ has the form $\exp[iae\Lambda_i/hc]$, so we may define magnetic translation operators

$$T_i = \exp\left[\frac{ia}{h}\left(p_i + \frac{e}{c}\Lambda_i\right)\right]. \tag{21}$$

The momentum operator generates the translation part, Λ_i generates the gauge transformation. If the magnetic field happens to be such that $\phi = m/n$ is a rational multiple, so we have commensurate flux, then $\left[T_x^q, T_y\right] = 0$. There is a subset of translations that commute with each other physically a rectangle made of q adjacent lattice unit cells contains an integer number of flux quanta. If that holds, we can simultaneously diagonalize H, T_x^q and T_y giving wave functions $\psi_{\mathbf{k}}$ labeled by $\mathbf{k} = (k_x, k_y)$ which satisfy

$$\begin{aligned}\psi_{\mathbf{k}}(x+qa,y) &= \exp\left[-iqa\left(\frac{k_x + eA_x}{hc}\right)\right]\psi_{\mathbf{k}}(x,y),\\ \psi_{\mathbf{k}}(x,y+a) &= \exp\left[-ia\left(\frac{k_y + eA_y}{hc}\right)\right]\psi_{\mathbf{k}}(x,y).\end{aligned} \tag{22}$$

These are the magnetic Block functions. The states are completely labeled by vector $\mathbf{k}$ together with a discrete band index. The vector $\mathbf{k}$ lies in the magnetic Brillouin zone.

From (16) (b), the phase change is given by $2\pi p$, with the wave function in the form

$$\varphi_{k_1,k_2}(x,y) = |\varphi_{k_1,k_2}(x,y)| \cdot \exp|i\theta_{k_1,k_2}(x,y)| \tag{23}$$

Let $\int ds$ denote a counterclockwise line integral around the boundary of the magnetic unit cell, then p is given by

$$p = -\frac{1}{2\pi}\int ds\,\frac{\partial\theta_{k_1k_2}(x,y)}{\partial s}. \tag{24}$$

In fact (20) is gauge invariant, although $\theta_{k_1,k_2}(x,y)$ depends on a gauge. This equation represents an important topological feature of the system. An arrow whose directional angles are given by θ_{k_1,k_2}, where the arrow goes around the boundary. Here, we have displayed a topological constraint that pertains to the wave function. It is remarked that Schrödinger equation (12) can be expressed in the following way as well,

$$H(k_1,k_2)\varphi^{\alpha}_{k_1k_2} = E^{\alpha}\varphi^{\alpha}_{k_1k_2}, \tag{25}$$

$$H(k_1,k_2) = \frac{1}{2m}(-i\hbar\nabla + \hbar\mathbf{k} + e\mathbf{A})^2 + U(x,y), \tag{26}$$

where $\mathbf{k} = (k_1,k_2)$ and E^{α} depends on $\mathbf{k}$.

4. Hall conductance

A small electric field can be applied to the system, which results in a current. This may be determined by the linear response or Kubo formula. The current density was introduced in (2) with (3). In equilibrium, a current in the direction requires the presence of an electric field with a component in the y-direction. The N-body Hamiltonian, in the absence of the external electric field is

$$\begin{aligned} H = \sum_{i=1}^{N}\frac{1}{2m_i}\left(-i\hbar\frac{\partial}{\partial x_i}\right)^2 + \frac{1}{2m_i}\left(-i\hbar\frac{\partial}{\partial y_i} - eBx_i\right)^2 + \sum_{i=1}^{N}U(x_i,y_i) \\ + \sum_{i=1}^{N}\sum_{\substack{i=1\\ i=j}}^{N}V(|\mathbf{r}_i - \mathbf{r}_j|). \end{aligned} \tag{27}$$

To write the Kubo formula we need two eigenvalues or energies and the velocity operators appearing in the Kubo formula. These are given by

$$v_1 = \sum_{i=1}^{N}\frac{1}{m_i}\left(-i\hbar\frac{\partial}{\partial x_i}\right), \qquad v_2 = \sum_{i=1}^{N}\frac{1}{m_i}\left(-i\hbar\frac{\partial}{\partial y_i} - eBx_i\right). \tag{28}$$

Realistic boundary conditions take the form $i = 1, \dots, N$,

$$\psi(x_i = L_1) = \psi(x_i = 0) = 0, \qquad \psi(y_i + L) = e^{i\beta L_2}\psi(y_i). \tag{29}$$

Linear response of a current in the perpendicular direction to the applied electric field is represented by the Hall conductance given by

$$\sigma_{xy} = -ie^2\hbar \sum_{E^{\alpha}<E_F<E^{\beta}} \frac{(v_y)_{\alpha\beta}(v_x)_{\beta\alpha} - (v_x)_{\alpha\beta}(v_y)_{\beta\alpha}}{(E^{\alpha} - E^{\beta})^2}. \tag{30}$$

In (30), E_F is the Fermi energy and the sum is over all states below and above the Fermi energy, so α, β label bands. A state is specified by a vector $\mathbf{k}$ as well as α. The matrix elements in (30) are of the velocity operator, which is

$$\mathbf{v} = \frac{1}{m}(-i\hbar\nabla + e\mathbf{A}) \tag{31}$$

obtained by integration over (x,y) for a single magnetic unit cell

$$(\mathbf{v})_{\alpha\beta} = \delta_{k_1',k_1}\delta_{k_2'k_2}\int_0^{qa}dx\int_0^{b}dy\,\varphi^{\alpha *}_{k_1k_2}\mathbf{v}\varphi^{\beta}_{k_1k_2}. \tag{32}$$

The states are normalized over the unit cell

$$\int_0^{qa} dx \int_0^b dy \, |\varphi^{\alpha}_{k_1 k_2}|^2 = 1. \tag{33}$$

The velocity operator in (29) can be replaced by partial derivatives of the k-dependent Hamiltonian (14), as only off-diagonal elements are considered

$$(v_x)_{\alpha\beta} = \frac{1}{\hbar}\left\langle \alpha | \frac{\partial H}{\partial k_1} | \beta \right\rangle, \qquad (v_y)_{\alpha\beta} = \frac{1}{\hbar}\left\langle \alpha | \frac{\partial H}{\partial k_2} | \beta \right\rangle. \tag{34}$$

The matrix elements of the partial derivatives of the Hamiltonian for $j = 1, 2$ are

$$\left\langle \alpha | \frac{\partial H}{\partial k_j} | \beta \right\rangle = \left(E^{\beta} - E^{\alpha}\right)\left\langle \varphi^{\alpha} | \frac{\partial \varphi^{\beta}}{\partial k_j} \right\rangle = -\left(E^{\beta} - E^{\alpha}\right)\left\langle \frac{\partial \varphi^{\alpha}}{\partial k_j} | \varphi^{\beta} \right\rangle. \tag{35}$$

Substitute (34), (35) into (30) to find that σ_{xy} is given by

$$\sigma_{xy} = -i\frac{e^2}{\hbar} \sum_{E^{\alpha} < E_F < E^{\beta}} \left(\left\langle \frac{\partial \varphi^{\alpha}}{\partial k_2} | \beta \right\rangle \left\langle \beta | \frac{\partial \varphi^{\alpha}}{\partial k_1} \right\rangle - \left\langle \frac{\partial \varphi^{\alpha}}{\partial k_1} | \beta \right\rangle \left\langle \beta | \frac{\partial \varphi^{\alpha}}{\partial k_2} \right\rangle \right). \tag{36}$$

Using a resolution of the identity $\sum(|\alpha\rangle\langle\alpha| + |\beta\rangle\langle\beta|) = 1$, σ^{α}_{xy} is obtained as a contribution to the Hall conductance from the completely filled band α,

$$\sigma^{\alpha}_{xy} = -i\frac{e^2}{\hbar}\cdot\frac{1}{2\pi}\int d^2k \int d^2r \left(\left\langle \frac{\partial \varphi^{\alpha}_{k_1 k_2}}{\partial k_2} | \frac{\partial \varphi^{\alpha}_{k_1 k_2}}{\partial k_1} \right\rangle - \left\langle \frac{\partial \varphi^{\alpha}_{k_1 k_2}}{\partial k_1} | \frac{\partial \varphi^{\alpha}_{k_1 k_2}}{\partial k_2} \right\rangle \right). \tag{37}$$

The derivatives in (37) are formal at this point and it still may not be seen why the Hall conductance should be quantized. It needs to be assumed that there is always a finite energy gap between the ground state and the excitations under any excitation boundary conditions. It is plausible the bulk conductance, as given by the Kubo formula, should be insensitive to the boundary conditions if the particles do not have long-range correlations in the ground state. In the special case of zero interaction and flat substrate potential, it can be shown explicitly that the above expression is indeed independent of the parameters, even without taking the thermodynamic limit.

Consequently σ can be equated with its average over the set of phases $\{0 \le \theta < 2\pi, 0 < \phi \le 2\pi\}$ that specify different boundary conditions

$$\sigma \equiv \overline{\sigma} = \frac{e^2}{h}\int_0^{2\pi} d\theta \int_0^{2\pi} d\phi \, \frac{1}{2\pi i}\left[\left\langle \frac{\partial \varphi^{\alpha}}{\partial \phi} | \frac{\partial \varphi^{\alpha}}{\partial \theta} \right\rangle - \left\langle \frac{\partial \varphi^{\alpha}}{\partial \theta} | \frac{\partial \varphi^{\alpha}}{\partial \phi} \right\rangle \right]. \tag{38}$$

The energy gap forces the ground state to go back to itself up to an overall phase factor as θ, ϕ change by 2π, unless the ground state is not uniquely determined by the boundary conditions. Then again, we arrive to the Hall conductance being quantized into an integer multiple of e^2/h. In fact (38) is actually a topological invariant. These integrals also have physical significance, as the stability of the quantization of the Hall conductance against various kinds of perturbations can be explained.

In physical terms, if a conductor with no extrinsic Berry curvature is subjected to a magnetic field, electrons near the boundary perform skipping orbits. They essentially roll along the boundary in a direction defined by the magnetic field. The skipping orbits persist no matter how the boundary is shaped and provide a single-conducing channel for currents in a *2D* electron system with a magnetic field and sufficiently high electron mobility. These orbits give rise to the quantum Hall effect.

The observation of the fractional quantum Hall effect can also be associated with a vanishing of the longitudinal conductance in the zero temperature limit. This suggests that a Fermi gap must also exist at the fractional fillings near which the effect is observed. The ground state is known to be liquid-like, so correlations between electrons decay rapidly as separation becomes large. We can continue to equate the Hall conductance to its average over all different boundary conditions. Since a non-degenerate ground state always leads to an integral quantization, degeneracy is required in order to explain fractional quantization. If d is the degree of degeneracy, $\{\phi_k\}$ an orthogonal basis spanning the ground state Hilbert space,

$$\sigma = \overline{\sigma} = \frac{e^2}{hd} \sum_{k=1}^{d} \int_0^{2\pi} d\theta \int_0^{2\pi} d\phi \, \frac{1}{2\pi i} \left[\left\langle \frac{\partial \phi_k}{\partial \phi} \middle| \frac{\partial \phi_k}{\partial \theta} \right\rangle - \left\langle \frac{\partial \phi_k}{\partial \theta} \middle| \frac{\partial \phi_k}{\partial \phi} \right\rangle \right]. \tag{39}$$

Unlike the non-degenerate case, the integral over θ, ϕ is no longer a topological invariant, since variation of θ, ϕ by 2π does not necessarily map the ground state back onto itself. However, the summation over the integral may still be a topological invariant.

The ϕ_k in (39) by a unitary transformation. If there are no other ground states other than those generated by the parent states, (39) can be written

$$\sigma = \overline{\sigma} = \frac{e^2}{hp} \int_0^{2\pi p} d\theta \int_0^{2\pi} d\phi \, \frac{1}{2\pi i} \left[\left\langle \frac{\partial \psi_1}{\partial \phi} \middle| \frac{\partial \phi_1}{\partial \theta} \right\rangle - \left\langle \frac{\partial \phi_1}{\partial \theta} \middle| \frac{\partial \phi_1}{\partial \phi} \right\rangle \right]. \tag{40}$$

In (40), the summation has been absorbed into the integration over the extended range $0 \le \theta \le p 2\pi$ because different ground states can be obtained from ϕ_1 by continuous variation of θ by $2\pi, 4\pi, \dots$. The extended zone $0 \le \theta \le p 2\pi$, $0 \le \phi \le 2\pi$ can be regarded as a torus, so ϕ is dragged back to itself up to an overall phase factor as θ changes by $2\pi p$ or φ by 2π. Hence (40) can be expressed concisely as

$$\sigma = \frac{e^2}{h} \cdot \frac{c}{p}, \tag{41}$$

where c is the integer is given by the integral. As in the non-degenerate case, we can continue to attach a topological meaning to this integer if the extended zone is regarded as the base manifold.

Returning to the former case, another approach can be given for this. A vector field can now be defined in the magnetic Brillouin zone as

$$\mathbf{A}(k_1, k_2) = \int d^2 r \, \varphi^*_{k_1 k_2} \nabla_{\mathbf{k}} \varphi_{k_1 k_2} = \langle \varphi_{k_1 k_2} | \nabla_{\mathbf{k}} | \varphi_{k_1 k_2} \rangle, \tag{42}$$

where $\nabla_{\mathbf{k}}$ is the vector operator

$$\nabla_{\mathbf{k}} = \left(\frac{\partial}{\partial k_1}, \frac{\partial}{\partial k_2} \right). \tag{43}$$

Since a contribution from only a single band is considered, the α is omitted. From (37) and (42), σ_{xy}^{α} is

$$\sigma_{xy}^{\alpha} = \frac{e^2}{h}\frac{1}{2\pi i}\int d^2k\,[\nabla_{\mathbf{k}} \times \mathbf{A}(k_1,k_2)]_3. \tag{44}$$

The square bracket in (44) denotes the third component, and the integration is over the region $0 \le k_1 \le 2\pi/qa$, $0 \le k_2 \le 2\pi/b$. This magnetic Brillouin zone is topologically a torus T^2 rather than a rectangle in k-space. The two points in k-space $k_1 = 0$, $2\pi/qa$ or $k_2 = 0, 2\pi/b$, are identified as the same point in k-space. Since the torus has no boundary, Stokes theorem as applied to (43) gives $\sigma_{xy}^{\alpha} = 0$ if the vector field $\mathbf{A}(k_1,k_2)$ were uniquely defined on the entire torus T^2.

A nonzero value for σ_{xy}^{α} is a consequence of nontrivial topology for $\mathbf{A}(k_1,k_2)$, which can only be constructed provided the global topology of the torus space is noncontractible. If $\varphi_{k_1k_2}^{(\alpha)}(x,y)$ satisfies Schrödinger equation (12), then so does

$$\varphi_{k_1k_2}^{(\alpha)}(x,y)e^{if(k_1,k_2)} \tag{45}$$

Here $f(k_1,k_2)$ is arbitrary, smooth function of k_1, k_2 and is independent of x, y. This induces a transformation

$$\varphi'_{k_1k_2}(x,y) = \varphi_{k_1k_2}(x,y)\cdot e^{if(k_1,k_2)}. \tag{46}$$

This is just an overall change of phase and any physical quantity must remain invariant under the transformation. By (42), $\mathbf{A}(k_1,k_2)$ must undergo a transformation

$$\mathbf{A}'(k_1,k_2) = \mathbf{A}(k_1,k_2) + i\nabla_{\mathbf{k}}f(k_1,k_2). \tag{47}$$

By (44) and (46), $\sigma_{xy}^{(\alpha)}$ is invariant under transformation (47).

A nontrivial topology arises when the phase of the wavefunction cannot be uniquely determined smoothly on the whole Brillouin zone. The transformation (46) implies the overall phase for each state vector can be selected in an arbitrary way by requiring a component of the state vector to be real. This is not sufficient to fix the phase over the entire zone since $\varphi_{k_1k_2}^{(\alpha)}(x^0,y^0)$ vanishes for some $\{k_1,k_2\}$.

For simplicity, suppose $\varphi_{k_1k_2}(x^{(0)},y^{(0)})$ vanishes only at one point (k_1^0,k_2^0) in the zone and divide T^2 into two parts T_1^2 and T_2^2 with $\left(k_1^{(0)},k_2^{(0)}\right) \in T_1^2$. A different convention can be used in T_2^2 so another component of $\varphi_{k_1k_2}(x^{(1)},y^{(1)}) = \langle x^{(1)},y^{(1)}|\varphi_{k_1k_2}\rangle$ is real and $(x^{(1)},y^{(1)})$ in T_1^2 are taken so that $\varphi_{k_1k_2}(x^{(1)},y^{(1)})$ does not vanish in T_1^2. The overall phase is uniquely determined entirely on T^2.

At the boundary ∂T of T_1^2 and T_2^2 there is a mismatch in the phase

$$|\varphi_{k_1k_2}^{II}\rangle = e^{i\chi(k_1,k_2)}\,|\varphi_{k_1k_2}^{I}\rangle. \tag{48}$$

Function $\chi(k_1,k_2)$ in (48) is smooth in terms of its arguments on ∂T. This nontrivial topology of $|\varphi_{k_1k_2}\rangle$ is transferred to that of the vector field $\mathbf{A}(k_1,k_2)$. Thus smooth vector fields $\mathbf{A}_I$ and A_{II} can be defined on the sets T_I^2 and T_{II}^2 by means of (46).

The phase difference of the state vector given in (48) implies the following relationship between $\mathbf{A}_I$ and $\mathbf{A}_{II}$ on the boundary

$$\mathbf{A}_{II}(k_1,k_2) = \mathbf{A}_I(k_1,k_2) + i\nabla\chi(k_1,k_2). \tag{49}$$

Applying Stokes theorem to T_I^2 and T_{II}^2, respectively, gives

$$\begin{aligned} \sigma_{xy}^{(\alpha)} &= \frac{e^2}{h}\frac{1}{2\pi i}\left(\int_{T_I^2} d^2k\,[\nabla\times\mathbf{A}_I(k_1,k_2)]_3 + \int_{T_{II}^2} d^2k\,[\nabla\times\mathbf{A}_{II}(k_1,k_2)]_3\right) \\ &= \frac{e^2}{h}\frac{1}{2\pi i}\int_{\partial T} d^2k(\mathbf{A}_I(k_1,k_2) - \mathbf{A}_{II}(k_1,k_2)). \end{aligned} \tag{50}$$

The line integral over the boundary has a sign change as the boundary has an opposite orientation with respect to T_I^2 and T_{II}^2. Using (45), it is found that

$$\begin{aligned} \sigma_{xy}^{(\alpha)} &= \frac{e^2}{\hbar}n, \\ n &= \frac{1}{2\pi}\int_{\partial T} d^2k\nabla_{\mathbf{k}}\chi(k_1,k_2). \end{aligned} \tag{51}$$

In (51), n must be an integer for each of the state vectors as they must fit together exactly when a full revolution around the boundary is carried out. This idea can be generalized to the case when the function $\varphi_{k_1k_2}^{(\alpha)}$ has more than one zero by introducing the idea of fiber bundles.

The self-rotation implied by a nonzero Chern number guarantees the existence of traveling edge states. Two-dimensional materials with nonzero Chern numbers have the same universal conductance even if no magnetic field is present.

5. Construction of a fiber bundle and Chern class

Since the state $\exp[if(k_1,k_2)]\,|\varphi_{k_1k_2}\rangle$ is normalized whenever state $|\varphi_{k_1k_2}\rangle$ is, it seems natural to consider a principal $U(1)$ bundle over T^2. It is known a chart for a torus is composed of four neighborhoods $\{U_i\}_{i=1}^4$, where each U_i is a subspace of $\mathbb{R}^2$. If four regions $\overline{U}_i$ are defined as

$$\begin{aligned} &\overline{U}_1 = \left\{(k_1,k_2)|0\le k_1<\frac{\pi}{qa}, 0<k_2<\frac{\pi}{b}\right\}, \quad \overline{U}_2 = \left\{(k_1,k_2)|\frac{\pi}{qa}<k_1<\frac{2\pi}{qa}, 0<k_2<\frac{\pi}{b}\right\}, \\ &\overline{U}_3 = \left\{(k_1,k_2)|\frac{\pi}{qa}<k_1<\frac{2\pi}{qa}, \frac{\pi}{b}<k_2<\frac{2\pi}{b}\right\} \quad \overline{U}_4 = \left\{(k_1,k_2)|0<k_1<\frac{\pi}{qa}, \frac{\pi}{b}<k_2<\frac{2\pi}{b}\right\}. \end{aligned} \tag{52}$$

Each U_i in (52) can be chosen to be a slightly larger region and so each completely includes the corresponding set $\overline{U}_i$. A principle $U(1)$ bundle is a topological space which is locally isomorphic to $U_i\times U(1)$ in each neighborhood U_i. The global topology of a specific fiber is determined by $|\varphi_{k_1k_2}\rangle$.

To construct the fiber bundle, take a component of the state vector $\varphi\left(x^{(0)}, y^{(0)}\right) = \left\langle x^{(0)} y^{(0)} | \varphi_{k_1 k_2} \right\rangle$ which does not vanish in the overlaps of the U_i. Since each U_i is contractible, it is possible to choose a convention for the phase such that the factor

$$\exp\left[i\theta^{(i)}(k_1, k_2)\right] = \frac{\varphi^{(i)}(x^0, y^0)}{|\varphi^{(i)}(x^0, y^0)|} \tag{53}$$

is smooth in each neighborhood U_i except at zeros of $\varphi^{(i)}_{k_1 k_2}\left(x^{(0)}, y^{(0)}\right)$. However, it is not possible in general to have a global phase convention valid over all the neighborhoods. A transition function Φ_{ij} is used in the overlap between two neighborhoods $U_i \cap U_j$,

$$\Phi_{ij} = \exp\left[i\theta^{(i)}(k_1, k_2) - i\theta^{(j)}(k_1, k_2)\right] = \exp\left[if^{(ij)}(k_1, k_2)\right]. \tag{54}$$

If we regard Φ_{ij} as a map $\Phi_{ij} : U(1) \to U(1)$, a principal $U(1)$ bundle over T^2 is completely specified by the transition function, and a nontrivial fiber bundle has been constructed.

Fiber bundles can be classified by means of certain integers which characterize the transition functions. When a connection is put on the bundles, these integers correspond to integrals which involve a bundle curvature. It is usual to write a connection one form in terms of a one-form $\mathcal{A}$,

$$\omega = g^{-1}\mathcal{A}g + g^{-1}dg = \mathcal{A} + i\,d\chi. \tag{55}$$

Here $g = e^{i\chi} \in U(1)$ is a fiber and $\mathcal{A}$ a one form. The one form $\mathcal{A}$ is specified in terms of coordinates by defining

$$\mathcal{A}(k_1, k_2) = \tilde{A}_\mu(k_1, k_2)\,dk_\mu = \left\langle \varphi_{k_1 k_2} | \frac{\partial}{\partial k_\mu} | \varphi_{k_1 k_2} \right\rangle dk_\mu. \tag{56}$$

The transition functions Φ_{ij} act on fibers by left multiplication. Where two neighborhoods U_i and U_j overlap, a transition function $\Psi = \Psi_{ij}$ relates the local fiber coordinates g and g' in U_j and U_i in the following way

$$g' = \Psi g. \tag{57}$$

This is equivalent to the gauge transformation (46) and so using (53) and (57)

$$\mathcal{A}' = \Phi\mathcal{A}\Phi^{-1} + \Phi\, d\,\Phi^{-1} = \mathcal{A} - i\frac{\partial f}{\partial k_\mu}\,dk_\mu. \tag{58}$$

In fact, ω is invariant under (57) and (58) and is a connection one form with A given by (56). Once a connection is specified, there is a geometry on the topological space such that the curvature is given by the exterior derivative of A,

$$F = d\mathcal{A} = \frac{\partial \tilde{A}_\mu}{\partial k_\mu}\,dk_\nu \wedge dk_\mu. \tag{59}$$

The first Chern form is defined to be,

$$c_1 = \frac{i}{2\pi} F. \tag{60}$$

Integrating c_1 over the entire manifold the first Chern number C_1 is obtained as

$$C_1 = \frac{i}{2\pi} \int F = \frac{i}{2\pi} \int d\mathcal{A} = \frac{i}{2\pi} \int \frac{\partial \tilde{A}_\mu}{\partial k_\nu} dk_\nu \wedge dk_\mu. \tag{61}$$

The Chern number C_1 is an integer which is independent of the particular connection which is used and although framed in terms of T^2 holds for any smooth manifold. It is specified by the topology of this $U(1)$ principal bundle which is constructed based on the state vector. The Chern number remains a constant integer with respect to smooth changes in the bulk energy band structure. This holds provided topological phase transitions do not occur.

Comparing this with (44) gives

$$\sigma_{xy}^{(\alpha)} = -\frac{e^2}{\hbar} C_1. \tag{62}$$

A contribution to the Hall conductance from a single filled band is given by minus the first Chern number corresponding to (60) in units of $e^2/\hbar$.

6. Homotopy phases and topological invariants

The appearance of topology in this physical phenomena can be approached in another way. It is known that various magnetic phases which corresponds to the breaking of rotational symmetry are driven by binary interaction of electrons in spin Heisenberg or Ising systems with magnetization as an order parameter displaying electron correlations in the system. There are also multi-electron correlated states also driven by binary interaction, but do not correspond to symmetry breaking and not assigned any local order parameter. To such classes of topologically correlated states belong phases observed experimentally in 2D electron magnetic fields and referred to as quantum Hall states. These phases can be characterized by topological invariants which manifest system organization induced by the interaction. These phases tend to display uniform behavior of all electrons in the system with constraints imposed on all electrons and driven by interaction which results in long range quantum entanglement of all electrons simultaneously in the system.

The fractional quantum Hall effect is induced by the Coulomb interaction of electrons in 2D multi-particle systems exposed to a perpendicular magnetic field B strong enough that Landau level filling is fractional $\nu = N/N_0$ and belongs to specific hierarchies of fractions. The hierarchy is universal and independent of microscopic structure of the 2D system. When $\nu = 1, 2, \ldots$, the integer quantum Hall effect for 2D electrons is observed with plateaus of quantized Hall resistivity and vanishing longitudinal resistivity.

The main point of the homotopy approach to FQHE conditions is commensurability of cyclotron braids with an electron distribution defined by the electron Coulomb repulsion. When there is no magnetic field the sizes of the braids are arbitrary. They may be large or small without any constraints. In the presence of a magnetic field, interactions change. Braids for electrons in 2D acquire a size set by the size of the cyclotron orbit. The braid trajectories that participate in particle exchanges are of finite size as well.

Correlations of electrons in 2D can be described by the fundamental homotopy group in multi-particle systems. These are the so called braid groups [30]. Braid groups collect non-homotopic classes of multi-particle trajectories describing classical exchanges of electrons in their multi-particle coordinate spaces. Nonhomotopic trajectories cannot be transformed into each other by any continuous deformation without cutting. As electrons are indistinguishable, braids are closed loops in a multi-particle configuration space.

Braids defining electron exchange in the presence of a magnetic field must be built of cyclotron orbit pieces. These are of finite size in 2D. Cyclotron orbits in 2D are set by the magnetic flux quantum regardless of interaction. The commensurability of braid size with the electron distribution selects homotopy patterns. These are topological invariants which define the observed FQHE hierarchy. The drift movement along the magnetic field direction in 3D makes the cyclotron range of an arbitrary size in three dimensions. The elementary braids, the generators of the full braid group are called σ_i, $i = 1, \dots, N-1$, which are exchanges of the i-th electron with the $(i+1)$-electron at an arbitrary but fixed enumeration of electrons in the plane. It is always possible to enumerate electrons on a plane in such a way that the consecutive ones are the closest ones in the Wigner crystal.

One-dimensional representations of the full braid group, that is the homotopy group π_1 of the configuration space for N indistinguishable identical particles, define the weights for path integrals over trajectories. If the trajectories belong to separate homotopy classes, then defining a measure on the whole space of trajectories is not possible because of discontinuity. The sum should be done over all topologically nonequivalent classes with unitary weight factors. Nonequivalent trajectory classes can be formed by attaching additional closed multi-particle loops to open trajectories $\lambda_{a,b}$ which join points a and b in configuration space. The number of nonequivalent loops is the same as the number of elements in the full braid group, and they are all naturally nonhomotopic. All trajectories with loops fall into disjoint classes of nonholomorphic trajectories which cannot be unified or mixed by continuous deformations. These classes form together the complete domain of the Feynman path integral

$$I\left(z_1, \dots, z_N, t; z_1', \dots, z_N', t'\right) = \sum_{l \in \pi_1(\Omega)} e^{i\alpha_l} \int d\lambda_l e^{iS\left[\lambda_l\left(z_1, \dots, z_N, t; z_1', \dots z_N', t'\right)\right]/\hbar}. \tag{63}$$

In (63), $I\left(z_1, \dots, z_N, t; z_1', \dots, z_N', t'\right)$ is called the propagator [15, 25, 26]. It is the matrix element of the evolution operator of the total N-particle system in the position representation which determines the probability amplitude of a transition from $z_1, \dots, z_N, t$ to $z_1', \dots, z_N', t'$. The measure of the path integral is $d\lambda_l$ and the integral is taken over all possible classical trajectories joining points in the configuration space lying in the l-th sector of the trajectory space with the l-th braid loop adjoint. The weight factors $e^{i\alpha_l}$ contribute one-dimensional representations of the full braid group.

The permutation group S_N is the full braid group of N particles in $\mathbb{R}^n$ for $n \geq 3$. There are only two one-dimensional unitary representations $\sigma_i \to e^{i0} = 1$ and $\sigma_i \to e^{i\pi} = -1$ where σ_i stands for the group generators exchanging the i-th and $(i+1)$ particles, which correspond to bosons and fermions. For N particles in $\mathbb{R}^2$, the braid group is considerably more complicated than S_N and has an infinite number of one-dimensional unitary representations that are defined for generators of the braid group as $\sigma_j \to e^{i\theta_j}, j = 1, \dots, N-1, \theta_j \in (-\pi, \pi]$. Various values of θ_j correspond to various types of abelian anyons because the elements of the one-dimensional representation commute. Closed trajectories from the full braid group describe the exchanges of identical particles. Their one-dimensional unitary representations determine the particle statistics.

Composite anyons [27] including composite fermions can be referred to one-dimensional unitary representations of cyclotron subgroups of the full braid group generated by the generators for $p = 3,5,7,9, \dots,$ and $i = 1, \dots, N-1$ with q odd and $\alpha \in (-\pi, \pi]$

$$b_i^{(p)} = \sigma_i^p \to e^{iq\alpha}. \tag{64}$$

Every p corresponds to a cyclotron subgroup of type p, and hence to other types of composite particles that are associated with the representations. The σ_i, $i = 1, \dots, N$ are the generators of the initial full braid group. The group elements $b_i^{(p)}$ represent the new elementary exchange of the i-th particle with the $(i+1)$ particle with $(p-1)/2$ loops, which is clear by virtue of the definition of the previous elementary exchange σ_i. The $b_i^{(p)}$ generators do not satisfy the relation $\sigma_i\sigma_{i+1}\sigma_i = \sigma_{i+1}\sigma_i\sigma_{i+1}$ for $1 \leq i \leq N-2$ that is $b_i^{(p)}b_{i+1}^{(p)}b_i^{(p)} \neq b_{i+1}^{(p)}b_i^{(p)}b_{i+1}^{(p)}$, while the condition $\sigma_i\sigma_j = \sigma_j\sigma_i$ for $1 \leq i,j \leq N-1$, $|i-j| \geq 2$ does hold for $b_i^{(p)}$, that is $b_i^{(p)}b_j^{(p)} = b_j^{(p)}b_i^{(p)}$ [30].

The condition is not fulfilled for the cyclotron subgroup, its representations may generally depend on the index i. However unitary representations of the full braid group that are reduced to the cyclotron subgroup do not depend on the i, and the induced representations of the cyclotron subgroup of type p have the form as in (64). Let us propose associating the composite particles and Laughlin conditions with the one-dimensional unitary representations of cyclotron braid subgroups generated by (65) with different values of p for different types of composite particle.

To see how composite fermions can be given in terms of cyclotron groups, recall Jain's model of composite fermions. It assumes that every particle has a $p-1$ quanta flux tube attached to it, for p-type composite fermions oriented opposite to the external field. The degeneracy of each Landau level is $N_0 = SBe/hc$ neglecting spin and for the fractional filling ν, $N_0 = N/\nu$. The area of the system is S and the total flux of the external field B that passes through the area of the system is equal to SB. The equation

$$\frac{SB}{N} = \frac{hc}{e} \cdot \frac{1}{\nu} \tag{65}$$

gives $1/\nu$ flux quanta per particle, which is in accord with the previous estimate [14–16].

Then the weakening of the external field by the mean field of these local tubes takes place. This can be estimated by calculating the reduction of the external flux caused by the flux tubes

$$SB - N(p-1)\frac{hc}{e} = \pm SB'. \tag{66}$$

In (66), $\pm$ represents the possible orientation of the reduced resultant B' field that is along or opposite the external B field orientation. In a reduced magnetic field SB' the IQHE can take place so

$$SB'\left(\frac{hc}{e}\right)^{-1}=\frac{N}{n}, \tag{67}$$

where n indicates the number of the Landau level. The number of completely filled Landau levels all of which have the same degeneracy and depend on the strength of the field is

$$\frac{SB}{hc/e}=\frac{SBe}{hc}. \tag{68}$$

The initial, real external field corresponds to the fractional lowest Landau level filling

$$\nu=\frac{N}{N_0}=\frac{Nhc}{SBe}. \tag{69}$$

These relations can combine to allow us to write

$$\frac{Nhc}{\nu e}-N(p-1)\frac{hc}{e}=\frac{Nhc}{n}\left(\frac{1}{\nu}-(p-1)\right)=\pm\frac{Nhc}{ne}. \tag{70}$$

This simplifies to the form

$$\frac{1}{\nu}-(p-1)=\pm\frac{1}{n}. \tag{71}$$

Eq. (71) can be solved for ν,

$$\nu=\frac{1}{p-1\pm\frac{1}{n}} \tag{72}$$

This results in main sequence of ν fillings for the FQHE in a real external field expressed by the incompressible states of the full filling separated by gaps, if the Coulomb interaction does not reduce them, of subsequent n Landau levels in a reduced resultant field due to the $p-1$ flux tube quanta that are attached to the particles in the case of p-type composite fermions.

If there are k additional loops in the braid $\sigma_i^{(2k+1)}$ where $2k+1$ is the number of loops in the multi-cyclotron orbit, the larger effective magnetic field flux quantum for multi-loop orbits $(2k+1)h/e$ in the lowest Landau level. These larger effective flux quanta for multi-loop orbits define topological invariants. The new commensurability condition, now for σ_i^{2k+1} braids, instead of $S/N=h/eB_0$ is

$$\frac{S}{N}=(2k+1)\frac{h}{eB}. \tag{73}$$

This is fullfilled for $B=(2k+1)B_0$ so (73) can be written as $\nu=N/N_0=1/2k+1$. The braids σ_i^{2k+1} now take the role of braid group generators and the new braid group generated by them is a subgroup of the former full braid group. Then (73) can be put in an equivalent, more general form,

$$\frac{SB}{N} = \frac{h}{x_1 e} + \frac{h}{x_2 e} + \cdots + \frac{h}{x_q e}, \tag{74}$$

with $x_1, \ldots, x_q = 1$ to match (73) and $q = 2k + 1$, which is the number of loops in the multi-loop cyclotron orbit. This can be thought of as the decomposition of the total flux of the external magnetic field passing through the sample SB per particle and per each loop of the $q = 2k + 1$ loop cyclotron orbit [14–16].

It is natural to generalize (74) to include the nesting of particular loops of the multi-loop orbit to the next-nearest neighbor in the Wigner lattice. Assume $x_i > 1$ integers that indicate the ratio of points of next-nearest neighbors of selected order N' so $x_i = N/N'$ where N is the total number of electrons. Thus every x_i-th electron is matched by an i-th loop.

The pattern illustrates the nesting of every second electron for $n = 2$, that is, of only $N/2$ electrons participating in correlations and $SB/N = h/xe$. This example defines commensurability of every n-th electron with a single loop cyclotron orbit. The same may be applied to a multi-loop orbit, or each of its loops separately

$$\frac{SB}{N} = \frac{h}{x_1 e} \pm \frac{h}{x_2 e} \pm \cdots \pm \frac{1}{x_q e}. \tag{75}$$

The $x_i \geq 1$ in (75) are integers and $\pm$ indicates the possible orientation of the loop congruent or reverse of the figure-8 shape, with respect to the preceding loop, respectively. None of the loops is distinguished from the others, so the loops can be ordered such that $1 \leq x_1 \leq x_2 \leq \cdots \leq x_q$.

Let us discuss two specific cases, the first with $x_1 = x_2 = \ldots = x_{q-1} = x$ and $x_q = y$ and $\pm$ only before the last component, the second with $x_1 = x_2 = \ldots = x_{q-2} = x$ and $x_{q-1} = x_q = y$ and $\pm$ only before the last two. For the first, (75) can be written as the usual FQHE hierarchy formula

$$\frac{SB}{N} = (q-1)\frac{h}{xe} \pm \frac{h}{ye}. \tag{76}$$

This can be put in the form

$$e\frac{SB}{hN} = \frac{q-1}{x} \pm \frac{1}{y} = \frac{(q-1)y \pm x}{xy}. \tag{77}$$

Inverting this result, ν can be obtained by using $N_0 = SBe/h$ to obtain

$$\nu = \frac{N}{N_0} == \frac{xy}{(q-1)y \pm x}. \tag{78}$$

The particle-hole symmetry gives a mirrored hierarchy for Landau holes in the lowest Landau level for this case

$$\nu = 1 - \frac{xy}{(q-1)y \pm x}. \tag{79}$$

For the second case, (75) reduces to

$$e\frac{SB}{Nh} = (q-2)\frac{1}{x} \pm \frac{2}{y}. \tag{80}$$

Inverting (80), we have

$$\nu = \frac{(q-2)y \pm 2x}{xy} \tag{81}$$

The hierarchies of (77) and (78) explain all the fractions experimentally observed in the lowest Landau level of GaAs in two dimensions. This also includes the fractions $\nu = 5/13, 4/11, 3/8, 3/10, \ldots$ which cannot be attained by means of the conventional composite fermion models. These fractions correspond to $x > 1$, whereas the Jain hierarchy matches to the one generated by (78).

The quantum of magnetic field flux Φ defines in two-dimensions the size of a cyclotron orbit, Φ/B. The quantum of the magnetic field flux, Φ, varies with the homotopy phases. Due to rigid correlations in some homotopy phases, the lower value of magnetic flux quantum may not be attainable for trajectories, unless correlations are disrupted. Such behavior can be seen by looking at the quantization condition of Bohr-Sommerfeld. This links the area of the one-dimensional phase space covered by a closed classical phase trajectory with the corresponding number of quantum states. The usual formula is derived assuming the trajectory between turning points is loopless. For a different homotopy class in which only multi-loop trajectories are available, loopless trajectories are excluded by correlation constraints [15]. The rule, then is

$$\oint p_i d\mathbf{x} = (2k+1)n_i h \tag{82}$$

for a trajectory with additional k loops as the simplest when loopless trajectories are not available.

7. Summary

Strongly correlated electron systems in two dimensions, such as the Hall fluid, have been realized experimentally. A fascinating aspect of such systems is that they can have global and topological properties that cannot be easily dealt with using conventional condensed matter physics methods. The long-distance properties of the ground state and its low-lying excitations is described by a class of field theories called topological field theories. These kinds of systems have a type of order known as topological order. Such theories have also appeared in other areas of physics such as in string theory. The main task realized here has been to show at a basic level how topological ideas can be expressed in a geometric and or topological way and accounting for the flat plateaus of the quantum Hall effects that emerge from this description. Of particular significance is the description of the filling factors that occur in the FQHE by means of braid groups and braiding. It is likely that more new physics will emerge by further study of these kinds of systems this way.

Mathematics subject classification

81T13, 81T50, 81T20

Author details

Paul Bracken
Department of Mathematics, University of Texas, Edinburg, TX, USA

*Address all correspondence to: paul.bracken@utrgv.edu

References

[1] Thouless DJ. Topological Quantum Numbers in Nonrelativistic Physics. Singapore: World Scientific; 1998

[2] Wen X-G. Quantum Field Theory of Many-Body Systems. Oxford: Oxford UP; 2010

[3] Ryder L. Quantum Field Theory. Cambridge: Cambridge University Press; 1996

[4] Busso R. The holographic principle. Reviews of Modern Physics. 2002;**74**:825

[5] Bracken P. A model for the quantization of the hall resistance in the quantum Hall effect. Journal of Modern Physics. 2010;**1**:158

[6] Acquaviva G, Iorio A, Smaldone LT. Topologically inequivalent quantizations. Annals of Physics. 2021; **434**:168641

[7] Halperin B, Jain JK. The Fractional Quantum Hall Effect. Singapore: World Scientific; 2020

[8] Ezawa ZF. The Quantum Hall Effect. 3rd ed. Singapore: World Scientific; 2013

[9] Stone M. Quantum Hall Effects. Singapore: World Scientific; 1992

[10] Jain JK. Composite Fermions. Cambridge: Cambridge UP; 2007

[11] Bohm D. Note on a theorem of block concerning possible causes of superconductivity. Physics Review. 1949;**75**:502

[12] Mann H. Topological quantum numbers of n-particle systems. In: Symmetries in Science X. NY: Plenum Press; 1998. pp. 285-292

[13] Hofstadter DR. Energy levels and wave functions of block electrons in rational and irrational magnetic fields. Physical Review B. 1976;**14**:2239

[14] Jacak JE. Topological approach to electron correlations at fractional quantum Hall effect. Annals of Physics. 2021;**430**:168493

[15] Jacak JE. Application of path-integral quantization to indistinguishable particle systems topologically confined by a magnetic field. Physical Review A. 2018; **97**:012108

[16] Jacak JE. Forbidden trajectories for path integrals. Physical Review A. 2023; **107**:032207

[17] Iwai T. A bulk-edge correspondence between the second Chern number and the spectral flow. Annals of Physics. 2023;**457**:169436

[18] Huckestein B. Charged particles in random magnetic fields and the critical behavior in the fractional quantum Hall effect. Physical Review B. 1996;**53**: 3650

[19] Haldane FDM. Model of the quantum hall effect without Landau levels: Condensed matter realization of the parity anomaly. Physical Review Letters. 1988;**61**:2015

[20] Laughlin RB. Quantized motion of three two-dimensional electrons in a strong magnetic field. Physical Review B. 1983;**27**:3383

[21] Bracken P. Chiral anomaly in Euclidean $(2+1)$ dimensional space and an application to the quantum Hall effect. International Journal of Modern Physics B. 2008;**22**:2675

[22] Chern SS, Simons J. Characteristic forms and geoemetric invariants. Annals of Mathematics. 1974;**99**:48

[23] Zee A. Long distance physics of topological fluids. Progress of Theoretical Physics Supplement. 1992; **107**:77

[24] Karabali D, Nair VP. The role of the spin connection in quantum Hall effect: A perspective from geometric quantization. Physical Review D. 2016; **94**:064057

[25] Chaichian M, Demichev A. Path Integrals in Physics, Stochastic Processes and Quantum Mechanics. Vol. I. Bristol: IOP Publishing; 2001

[26] Chaichian M, Demichev A. Path Integrals in Physics, Quantum Field Theory, Statistical Physics and Other Modern Applications. Vol. II. Bristol: IOP Publishing; 2001

[27] Wilczek F. Fractional Statistics and Anyon Superconductivity. Singapore: World Scientific; 1990

[28] Klauder JR. Quantization is geometry after all. Annals of Physics. 1988;**188**:120

[29] Klauder JR, Onorfi E. Landau levels and geometric quantization. International Journal of Modern Physics A: Particles and Fields; Gravitation; Cosmology; Nuclear Physics. 1989;**4**: 3939

[30] Artin E. Theory of braids. Annals of Mathematics. 1947;**48**:101

Section 2

Application of Modern Topological Spaces

Chapter 5

Some Categorical Structures of Topological Fuzzes

Ghasem Mirhosseinkhani

Abstract

An object X of a category $\mathbf{A}$ with finite products is said to be exponentiable if the product functor $- \times X : \mathbf{A} \to \mathbf{A}$ is a co-adjoint, and $\mathbf{A}$ is said to be Cartesian closed, provided that every object of $\mathbf{A}$ is exponentiable. It is well known that the category **TopFuzz** of topological fuzz spaces is both complete and co-complete, but it is not Cartesian closed, so we study and introduce some characterizations of the exponentiable objects in this category.

Keywords: molecular lattice, fuzz, topological fuzz, exponential object, Cartesian closed category

1. Introduction

A completely distributive and complete lattice is called a molecular lattice. In 1992, Wang introduced the concept of topological molecular lattices in terms of closed elements as a generalization of general topological spaces, fuzzy topological spaces and L-fuzzy topological spaces. He presented concepts such as molecule, remote neighborhood and generalized order homomorphism for this purpose [1–3]. A fuzz is a pair (F') consisting of a molecular lattice F and an order reversing involution $' : F \to F$, that is, $x \leq y$ if and only if $y' \leq x'$ and $x'' = x$ for all $x, y \in F$. A topological fuzz (or fuzz space) is a triple (F', τ) such that (F') is a fuzz and $\tau \subseteq F$ is a topology, that is, it is closed under finite meets, arbitrary joins and $0, 1 \in \tau$, where 0 and 1 are the smallest and the greatest elements of F, respectively. Every element of a topology τ is called open and every element of τ' is called closed, where $\tau' = \{a' | a \in \tau\}$.

For two molecular lattices L_1 and L_2, and a mapping $f : L_1 \to L_2$ which preserves arbitrary joins, suppose $\hat{f}$ denote the right adjoint of f, then $\hat{f} : L_2 \to L_1$ is defined by $\hat{f}(y) = \vee\{x \in L_1 | f(x) \leq y|\}$ for every $y \in L_2$.

A map $f : L_1 \to L_2$ between molecular lattices is said to be a generalized order homomorphism or an **ml**-map if f preserves arbitrary joins and its right adjoint $\hat{f}$ preserves arbitrary joins and arbitrary meets. A map $f : (F_1{}') \to (F_2{}')$ between fuzzes is said to be an order homomorphism or a **fuzz**-map if f preserves arbitrary joins and its right adjoint $\hat{f}$ preserves $'$. It is easy to show that every **fuzz**-map is an **ml**-map, but the converse is not true, in general [2].

A **fuzz**-map $f : (F_1', \tau) \to (F_2', \mu)$ between fuzz spaces is said to be continuous if $\hat{f}(b) \in \tau$ whenever $b \in \mu$. Let **Fuzz** be the category whose objects are fuzzes and whose morphisms are **fuzz**-maps, and **TopFuzz** be the category whose objects are fuzz spaces and whose morphisms are continuous **fuzz**-maps. It is well known that these categories are both complete and co-complete, and some categorical structures of them were introduced by many authors [2, 4–10]. Fuzz spaces provide a more general framework for studying fuzzy topological spaces. Since the category **Top** of all topological spaces, as a full subcategory of **TopFuzz**, is reflective and co-reflective, this points out the essential difference between the general topological spaces and fuzz spaces on the categorical level.

In the following, we give the definition of extra order introduced in Ref. [11]. Extra orders are useful tools to construct molecular lattices and function spaces in topological molecular lattices and fuzz spaces. Recall that a partially ordered set, or poset for short, is a nonempty set P equipped with a partial order $\leq$.

Definition 1.1 [11] Let P be a poset and $\prec$ be a binary relation on P.

a. $\prec$ is called an extra order, if it satisfies the following conditions:

   1.$x \prec y \Rightarrow x \leq y$,

   2.$u \leq x \prec y \leq v \Rightarrow u \prec v$.

b. $\prec$ satisfies the interpolation property (short by INT), if $x \prec y$ implies that $z \in P$ such that $x \prec z \prec y$.

Remark 1.2 If $\prec$ is an extra order on a poset P, then there exists a largest extra order $\overline{\prec}$ over P contained in $\prec$ satisfying (INT).

Definition 1.3 [11] Let $\prec$ be an extra order satisfying the interpolation property on a poset P. A lower subset I of P is said to be a lower-Dedekind $\prec$-cut, if it satisfies the following condition:

For every $x \in I$ there is a $y \in I$ such that $x \prec y$.

Let L be a complete lattice L. Then we define an extra order $\lhd$ on L by: $a \lhd b$ if for every subset $S \subseteq L$, $b \leq \vee S$ implies $a \leq s$ for some $s \in S$. If L is a molecular lattice, then $\lhd$ satisfies the condition (INT). Moreover, a complete lattice L is a molecular lattice if and only if $b = \vee \lhd(b)$, where $\lhd(b) = \{a \in L | a \lhd b\}$.

The set of all lower-Dedekind $\prec$-cuts in P ordered by subset inclusion is denoted by $Low_{\prec}(P)$. The following important result is the construction of molecular lattices using extra order.

Theorem 1.4 [11] If $\prec$ is an extra order over P satisfying (INT), then $Low_{\prec}(P)$ is a molecular lattice. Moreover, for $I, J \in Low_{\prec}(P)$, $I \lhd J$ if and only if $I \subseteq \downarrow y$ for some $y \in J$.

For two fuzzes (F_1') and (F_2'), let $[F_1 \to F_2]$ denote the set of all continuous sup-preserving maps from F_1 to F_2. An extra order $\prec$ over $[F_1 \to F_2]$ is defined by $f \prec g$ if $x \lhd y$ implies $f(x) \lhd g(y)$.

Definition 1.5 [10] Let $f, g \in [F_1 \to F_2]$. Then f and g are called parallel and denoted by $f \| g$, if for every $x \in F_1, \hat{f}((g(x))') \leq x'$.

Theorem 1.6 [10] Let $[F_2^{F_1}] := Low_{\overline{\prec}}[F_1 \to F_2]$, and $' : [F_2^{F_1}] \to [F_2^{F_1}]$ is defined by $A' = \cup\{B \in [F_2^{F_1}] : f \nparallel g \text{ for all } f \in A \text{ and } g \in B\}$ for every $A \in [F_2^{F_1}]$. Then $\left([F_2^{F_1}]'\right)$ is a fuzz.

Remark 1.7 [10, 12] The product of two fuzz spaces (F_1', τ_1) and (F_2', τ_2) is $(F_1 \otimes F_2, ', \tau)$, where $F_1 \otimes F_2 := \{D \subseteq F_1 \times F_2 | D = \cup_{(x,y) \in D} \lhd(x) \times \lhd(y)\}$, $D' = \cap_{(x,y) \in D}\{(\lhd(x') \times \lhd(1)) \cup (\lhd(1) \times \lhd(y'))\}$ for each $D \in F_1 \otimes F_2$ and τ is generated by subbase $\{\hat{\pi}_1(x) | x \in \tau_1\} \cup \{\hat{\pi}_2(y) | y \in \tau_2\}$, such that the projection **Fuzz**-maps π_1 and π_2 are defined by $\pi_1(D) = \vee\{x \in F_1 | \exists y \in F_2, (x,y) \in D\}$ and $\pi_2(D) = \vee\{y \in F_2 | \exists x \in F_1, (x,y) \in D\}$.

Definition 1.8 [1] An element a of a lattice L is said to be co-prime or molecule, if the following condition holds:

$$a \leq b \vee c \Rightarrow a \leq b \text{ or } a \leq c. \quad (1)$$

The set of all nonzero co-prime elements is denoted by $Cop(L)$. If L is a molecular lattice, then every element of L is a join of some elements of $Cop(L)$.

2. Exponentiable objects in TopFuzz

An object X of a category **A** with finite products is said to be exponentiable if the product functor $X \times - : \mathbf{A} \to \mathbf{A}$ is a co-adjoint, and it is called Cartesian closed, provided that every object of **A** is exponentiable [13]. Thus, an object X is exponentiable if and only if for each object Y there exists an object Y^X and a morphism $ev : X \times Y^X \to Y$ with the following universal property: for each morphism, $f : X \times Z \to Y$ there exists a unique morphism $\hat{f} : Z \to Y^X$ such that $f = ev \circ \left(id_X \times \hat{f}\right)$. The object Y^X is called the power object and the morphism ev is called the evaluation morphism. For example, in the category **Set** of sets we have the power object Y^X is the set of all functions from X to Y, and the evaluation morphism $ev : X \times Y^X \to Y$ is defined by $ev(x,f) = f(x)$. Thus **Set** is a Cartesian closed category.

Since $X \times - : \mathbf{Top} \to \mathbf{Top}$ does not preserve quotients, it follows that the category **Top** is not Cartesian closed. There are many characterizations of the exponentiable objects in **Top**, one interesting characterization is that a topological space is exponentiable if and only if it is core compact, that is, for every $x \in X$ and every open neighborhood V of x, there exists an open neighborhood U of x such that every open cover of V has a finite subcover of U [14, 15]. Similar results for the exponentiable objects in the category **Tml** of topological molecular lattices were introduced in Refs. [4, 16]. In this section, we introduce some characterizations of exponentiable objects in **TopFuzz**.

Let (F', τ) be a fuzz space. We define a relation $\ll$ on F as follows:

$a \ll b$ if and only if for every subset U of τ, $b \leq \vee U$ implies that there is a finite subset O of U such that $a \leq \vee O$.

Definition 1.9 A fuzz space (F', τ) is said to be core compact if for every $a \in \tau$, $a = \vee\{b \in \tau | b \ll a\}$.

Example 1.10 A topological space (X, τ) is core compact if and only if $(\rho(X), \tau)$ is a core compact fuzz space, where $\rho(X)$ is the power set of X and involution on $\rho(X)$ is the subset complement.

Definition 1.11 Let (F', τ) be a fuzz space. An upper subset H of τ is said to be Scott open if the following condition holds:

For every subset U of τ, $\vee U \in H$ implies that there is a finite subset O of U such that $\vee O \in H$.

The set of all Scott open sets on τ written by $\sigma(\tau)$. Let (F_1', τ_1), (F_2', τ_2) be two fuzz spaces, $H \in \sigma(\tau_1)$, $x \in \tau_2$ and $T(H,x) = \wedge\left\{E \in F_2^{F_1} | \hat{f}(x) \in H \text{ for every} f \in E\right\}$. The topology on $F_2^{F_1}$ is called Isbell topology and generated by $\{T(H,x) | H \in \sigma(\tau_1), x \in \tau_2\}$. In the following we will denote it by η.

Lemma 1.12 Let (M', τ), (F_1', τ_1) and (F_2', τ_2) be fuzz spaces and $F : M \otimes F_1 \to F_2$ be a continuous **fuzz**-map. Then for every $p \in Cop(M)$, the mapping $f_p : F_1 \to F_2$ that defines as $f_p(a) = F(\lhd(p) \times \lhd(a))$, is a continuous sup-preserving map.

Proof: f_p is a sup-preserving map, because F and $\lhd$ preserve supremum.

Now our goal is to show that it is continuous. Let $b \in \tau_2$. Then

$$\begin{aligned} \hat{f}_p(b) &= \bigvee\left\{a \in M | f_p(a) \leq b|\right\} \\ &= \bigvee\{a \in M | a \in Cop(M) \quad \text{and} \quad \lhd(p) \times \lhd(a) \leq \hat{F}(b)\} \end{aligned} \tag{2}$$

Since $\hat{F}(b)$ is open in $M \otimes F_1$, there exists a set I, an $x_i \in \tau$ and $y_i \in \tau_1$ for each $i \in I$ such that $F(\hat{b}) = \vee_{i \in I} \hat{\pi}_M(x_i) \wedge \hat{\pi}_{F_1}(y_i,)$. In the following we will prove $\hat{f}_p(b) = \vee_{p \leq x_i} y_i$ that shows $\hat{f}_p(b)$ is open and thus f_p is continuous. Let $\lhd(p) \times \lhd(a) \leq \hat{F}(b)$ for some $a \in Cop(M)$. Then $p \leq x_i$ and $a \leq y_i$ for some $i \in I$. Therefore $a \leq \vee_{p \leq x_i} y_i$ and finally $\hat{f}_p(b) \leq \vee_{p \leq x_i} y_i$. To show that $\vee_{p \leq x_i} y_i \leq \hat{f}_p(b)$, it suffices to check for every $i \in I$ that $p \leq x_i, y_i \leq \hat{f}_p(b)$. Let $p \leq x_i$. Then $\lhd(p) \times \lhd(y_i) \leq \lhd(x_i) \times \lhd(y_i) \leq \hat{\pi}_M(x_i) \wedge \hat{\pi}_{F_1}(y_i) \leq \vee_{i \in I} \hat{\pi}_M(x_i) \wedge \hat{\pi}_{F_1}(y_i) = \hat{F}(b)$. So for every $i \in I$ that $p \leq x_i, y_i \leq \hat{f}_p(b)$ and hence, $\vee_{p \leq x_i} y_i \leq \hat{f}_p(b)$. Therefore $\vee_{p \leq x_i} y_i = \hat{f}_p(b)$.

Lemma 1.13 Let $F : M \otimes F_1 \to F_2$ be a continuous **fuzz**-map. Then the mapping $\Lambda(F) : M \to F_2^{F_1}$ defined by $\Lambda(F)(a) = \downarrow\left\{f_p | p \lhd a \text{ and } p \in Cop(M)\right\}$ is a continuous **fuzz**-map.

Proof: It is easy to prove that $\Lambda(F)$ is a **fuzz**-map. So, it remains to show that $\Lambda(F)$ is continuous with respect to τ and η. For this goal it suffices to verify that $\hat{\Lambda}(F)(T(H,x)) \in \tau$, for every $x \in \tau_2$ and $H \in \sigma(\tau_1)$. Let $\hat{\Lambda}(F)(T(H,x)) = b$ and $A = \{a \in M | \Lambda(F)(a) \leq T(H,x)\}$. Then $b = \vee A$. In the following to show that $b \in \tau$, we will prove that for every $p \in Cop(M)$ if $p \leq b$, then there exists an open object $O \in \tau$ such that $p \leq O \leq b$. Let $p \leq b$. Then there exists an object $a_0 \in A$ such that $p \leq a_0$ and $\Lambda(F)(p) \leq \Lambda(F)(a_0) \leq T(H,x)$. Hence for every co-prime element $c \lhd p$ we have $\hat{F}(\lhd(c) \times \lhd())(x) \in H$. Let $y = \hat{f}_p(x)$. Then $y \in H$ and $y = \vee\left\{l \in F_1 | f_p(l) \leq x|\right\} = \vee\{l \in F_1 | \lhd(p) \times \lhd(l) \leq \hat{F}(x)\}$.

As $\hat{F}(x)$ is an open object in $(M, \tau) \otimes (F_1, \tau_1)$, there exists a set I, an $x_i \in \tau$ and $y_i \in \tau_1$ for each $i \in I$ such that $\hat{F}(x) = \vee \hat{\pi}_M(x_i) \wedge \hat{\pi}_{F_1}(y_i)$. By the proof of the last Lemma, $y = \hat{f}_p(x) = \vee_{p \leq x_i} y_i \in H$. Since $H \in \sigma(\tau_1)$, there exist finite elements $x_1, x_2, \dots, x_n$ such that $p \leq x_i$ for every $i = 1, \dots, n$ and $\tilde{y} = (y_1 \vee \dots \vee y_n) \in H$. Let $O = x_1 \wedge \dots \wedge x_n$. Then $O \in \tau$ and it is trivial that $p \leq O$.

Now we show that $O \leq b$. Let $m \lhd O$. So $m \leq x_i$ for each $i = 1, \dots, n$ and it follows that $\lhd(m) \times \lhd(\tilde{y}) \leq \vee \hat{\pi}_M(x_i) \wedge \hat{\pi}_{F_1}(y_i) = \hat{F}(x)$ and $f_m(\tilde{y}) \leq x$. So $\tilde{y} \leq \hat{f}_m(x)$. H is an upper subset of τ_1 and $\tilde{y} \in H$, so $\hat{f}_m(x) \in H$. If $f \in \Lambda(F(m))$, then $f \leq f_p$ for some

$p \lhd m$. Since $\hat{f}_m \leq \hat{f}_p$, it follows that $\hat{f}(x) \in H$. Thus $\Lambda(F(m)) \subseteq T(H,x)$, and hence $m \leq \hat{\Lambda}(F)(T(H,x)) = b$. Therefore, we have $O \leq b$, as desired.

Lemma 1.14 For every core compact fuzz space (M', τ), the evaluation map $ev : (F_1^M, \eta) \otimes (M, \tau) \to (F_1, \tau_1)$ is a continuous **fuzz**-map.

Proof: It is easy to show that ev is a **fuzz**-map. In the following to show that ev is continuous, we prove that $\hat{ev}(x)$ is open in $(F_1^M, \eta) \otimes (M, \tau)$ for every $x \in \tau_1$. So it is enough to check that for any $D \in Cop(F_1^M)$ and $p \in Cop(M)$, if $\lhd(D) \times \lhd(p) \leq \hat{ev}(x)$, then there exists an open object O of $F_1^M \otimes M$ such that $\lhd(D) \times \lhd(p) \leq O \leq \hat{ev}(x)$, because the set $\{\lhd(D) \times \lhd(p) | D \in Cop(F_1^M), p \in Cop(M)\}$ is a join generating base for $F_1^M \otimes M$.

Supposing that $\lhd(D) \times \lhd(p) \leq \hat{ev}(x)$ for some $D \in Cop(F_1^M)$ and $p \in Cop(M)$, then for any $f \in D$ we have $f(p) \leq x$, that is, $p \leq \hat{f}(x)$. Since τ is a continuous lattice, there exists $\tilde{l} \in \tau$ such that $p \leq \tilde{l}$ and $\tilde{l} \ll \hat{f}(x)$. Let $H = \{F_1 \in \tau | \tilde{l} \ll F_1\}$. Then $H \in \sigma(\tau)$, and hence $T(H,x) \in \eta$. While $f \in D$ and $\hat{f}(x) \in H$, it follows that $D \leq T(H,x)$. Therefore $\lhd(D) \times \lhd(p) \leq \hat{\pi}_{F_1^M}(T(H,x)) \wedge \hat{\pi}_M(\tilde{l})$. If $O = \hat{\pi}_{F_1^M}(T(H,x)) \wedge \hat{\pi}_M(\tilde{l})$, then O is open in $F_1^M \otimes L$. Now we show that $O \leq \hat{ev}(x)$. Let $D_0 \in Cop(F_1^M)$ and $p_0 \in Cop(M)$ and $\lhd(D_0) \times \lhd(p_0) \leq O$. Then $D_0 \leq T(H,x)$ and $p_0 \leq \tilde{l}$. So for every $g \in D_0$ it follows that $\tilde{l} \leq \hat{g}(x)$ and $g(p') \leq x$. Therefore $sup\{h(p_0) | h \in D_0\} \leq x$. On the other hand, it is obvious that $ev(\lhd(D_0) \times \lhd(p_0)) = sup\{h(p_0); h \in D'\}$. So $\lhd(D_0) \times \lhd(p_0) \leq \hat{ev}(x)$ when $\lhd(D_0) \times \lhd(p_0) \leq O$ for every $D_0 \in Cop(F_1^M)$ and $p_0 \in Cop(M)$, hence $O \leq \hat{ev}(x)$.

Lemma 1.15 Let (M', τ) be a core compact fuzz space and (F_1', τ_1), (F_2', τ_2) be arbitrary fuzz spaces. For every continuous **fuzz**-map $F : F_1 \otimes M \to F_2$, $\Lambda(F) : F_1 \to F_2^M$ is the unique continuous **fuzz**-map with respect to τ_1 and η such that $evo(\Lambda(F) \otimes id) = F : F_1 \otimes M \to F_2^M \otimes M \to F_2$.

Proof: Let $\Lambda_1(F), \Lambda_2(F) : F_1 \to F_2^M$ be arbitrary **fuzz**-maps such that $ev\circ(\Lambda_1(F) \otimes id) = ev\circ(\Lambda_2(F) \otimes id) = F$. We show that $\Lambda_1(F) = \Lambda_2(F)$. Since $\Lambda_1(F)$ and $\Lambda_2(F)$ preserve arbitrary joins, if $a \lhd b$, then $\Lambda_1(F)(a) \lhd \Lambda_2(F)(b)$ for every $a, b \in F_1$. Therefore by Theorem 1.4, there exists an $h \in \Lambda_1(F)(b)$ such that $\Lambda_1(F)(a) \subset \downarrow h$. So for every $x \in M$ we have:

$$\begin{aligned} ev\circ(\Lambda_2(F) \otimes id)(\lhd(a) \times \lhd(x)) &= ev\circ(\Lambda_1(F) \otimes id)(\lhd(a) \times \lhd(x)) \\ &= \vee\{k(x) | k \in \Lambda_1(F)(x)\} \leq h(x) \end{aligned} \tag{3}$$

Thus $\Lambda_2(F)(a) \subseteq \downarrow h \subseteq \Lambda_1(F)(b)$ and hence $\Lambda_2(F) \leq \Lambda_1(F)$. Similarly $\Lambda_1(F) \leq \Lambda_2(F)$. So $\Lambda_1(F) = \Lambda_2(F)$.

Lemma 1.16 $(F_1', \tau_1) \rightrightarrows (F_2', \tau_2) \xrightarrow{e} (F_3', \tau_3)$ is a co-equalizer in **TopFuzz** if and only if $F_1 \rightrightarrows F_2 \xrightarrow{e} F_3$ is a co-equalizer in the category **Mol** of molecular lattices and $\tau_3 = \{a \in F_3 | \hat{e}(a) \in \tau_2\}$.

Corollary 1.17 A **fuzz**-map $e : (F_1', \tau_1) \to (F_2', \tau_2)$ is a regular epimorphism in **Fuzz** if and only if $e : F_1 \to F_2$ is a regular epimorphism in **Mol** and $\tau_2 = \{a \in F_2 | \hat{e}(a) \in \tau_1\}$.

Lemma 1.18 A fuzz space (M', τ) is core compact if and only if for every $a \in \tau$, $p \in Cop(M)$, if $p \leq a$, then there exist an $H \in \sigma(\tau)$ and $b \in \tau$ such that $a \in H$, $p \leq b$ and $b \leq \wedge H$.

Proof: Let (M', τ) be a core compact fuzz space and $p \leq a$ for $p \in Cop(M)$, $a \in \tau$. Then there exists an $H \in \sigma(\tau)$ and $b \in \tau$ such that $p \leq b$, $b \ll a$. Supposing that $H = \{a \in M | b \ll a\}$. Then $H \in \sigma(\tau)$, $a \in H$ and $b \leq \wedge H$.

For converse let $a \in \tau$. It is obvious that $\vee\{b \in \tau | b \ll a\} \leq a$. Now assume that $p \in Cop(M)$ is an arbitrary object that $p \leq a$ and there exists a $b \in \tau$ such that $a \in H$, $p \leq b$ and $b \leq \wedge H$. We will show that $b \ll a$. Since H is an upper subset of τ, if $a \leq \vee_{i \in I} a_i$, then $\vee_{i \in I} a_i \in H$. Therefore there exist finite objects $a_1, \dots, a_n$ that $a_1 \vee \dots \vee a_n \in H$. So $b \leq \wedge H \leq a_1 \vee \dots \vee a_n$ and it means that $b \ll a$. In fact it follows that if $p \leq a$ then $p \leq \vee\{b | b \ll a|\}$. Hence $a \leq \vee\{b | b \ll a|\}$ and equality follows.

Lemma 1.19 Let (M', τ) be a fuzz space and $(\rho(\tau), \sigma(\tau))$ a usual topological space. Then $T = \cup\{\{a\} \times \lhd(p) | p \in Cop(M), a \in \tau, p \leq a\}$ is open in $(\rho(\tau), \sigma(\tau)) \otimes (M, \tau)$ if and only if (M', τ) is core compact.

Proof: It is clear that $T \in \rho \otimes L$. Because $T = \cup_{(a,p) \in T} \lhd(a) \times \lhd(p)$.

Now let T be open in $(\rho(\tau), \sigma(\tau)) \otimes (M, \tau)$, $a \in \tau$ and $p \in Cop(M)$ such that $p \leq a$. According to the previous Lemma, if there exists an $H \in \sigma(\tau)$ and $b \in \tau$ such that $a \in H$, $p \leq b$ and $b \leq \wedge H$, then (M, τ) is core compact. So we should find H and b. Since $p \leq a$ and T is open, then $\{a\} \times \lhd(p) \leq \cup\{\{a\} \times \lhd(p,) | p \in Cop(M), a \in \tau, p \leq a\}$ and there exists an $H \in \sigma(\tau)$ and $b \in \tau$ such that $\{a\} \times \lhd(p) \leq \hat{\pi}_\tau(H) \wedge \hat{\pi}_M(b)$ and $\hat{\pi}_\tau(H) \wedge \hat{\pi}_M(b) \leq T$. The first inequality shows that $a \in H$ and $p \leq b$. So, it remains to prove $b \leq \wedge H$. Let $p_1 \in Cop(M)$, $a_1 \in H$ such that $p_1 \leq b$. Then $\{a_1\} \times \lhd(p_1) \leq \hat{\pi}_\tau(H) \wedge \hat{\pi}_M(b) \leq T$. Therefore $p_1 \leq a_1$, whenever $p_1 \leq b$ for every $a_1 \in H$. That is $b \leq \wedge H$.

For the sufficiently, let (M, τ) be a core compact fuzz space. If $\{a\} \times \lhd(p) \leq T$ for every $a \in \tau$ and $p \in Cop(M)$, then $p \leq a$. By the previous Lemma, there exists an $H \in \sigma(\tau)$ and $b \in \tau$ such that $a \in H$, $p \leq b$ and $b \leq \wedge H$. In the following, we show that $\{a\} \times \lhd(p) \leq \hat{\pi}_\tau(H) \wedge \hat{\pi}_M(b)$ and $\hat{\pi}_\tau(H) \wedge \hat{\pi}_M(b) \leq T$. Since $a \in H$ and $p \leq b$ the first inequality is obvious. Now supposing that $a_1 \in \tau$ and p_1 is a co-prime element of M that $\{a_1\} \times \lhd(p_1) \leq \hat{\pi}_\tau(H) \wedge \hat{\pi}_M(b)$. So $p_1 \leq a_1$ and $\{a_1\} \times \lhd(p_1) \leq T$ and it follows that $\hat{\pi}_\tau(H) \wedge \hat{\pi}_M(b) \leq T$. Hence T is open in $(\rho(\tau), \sigma(\tau)) \otimes (M, \tau)$.

Definition 1.20 Let (M', τ) be a fuzz space. φ is called an admissible topology of τ, if T is open in $(\rho(\tau), \varphi) \otimes (M, \tau)$.

Lemma 1.21 Let (M', τ) be a fuzz space. Then $\sigma(\tau)$ is the infimum of the set of admissible topologies, i.e., $\sigma(\tau) = \cap\{\varphi : \varphi \text{ is an admissible topology of } \tau\}$.

Proof: Supposing that T is open in $(\rho(\tau), \varphi) \otimes (M, \tau)$, in the following we prove that $\sigma(\tau) \subseteq \varphi$. Let $H \in \sigma(\tau)$, we will show that $H \in \varphi$ and for this goal, it is enough to check that if $a \in H$ for every $a \in \tau$, then there exists an $\tilde{H} \in \varphi$ such that $a \in \tilde{H}$ and $\tilde{H} \subseteq H$.

Now let $a \in H$ and $p \leq a$ for every $p \in Cop(M)$. Then $\{a\} \times \lhd(p) \leq T$. Since T is open in $(\rho(\tau), \varphi) \otimes (M, \tau)$, there exists an $H_p \in \varphi$ and $a_p \in \tau$ such that $\{a\} \times \lhd(p) \leq \hat{\pi}_\tau(H_p) \wedge \hat{\pi}_M(a_p) \leq T$. Therefore $a \in H_p$ and $p \leq a_p$ for every $p \in Cop(M)$, that is, $a \leq \vee_{p \leq a} a_p$. In the following we show that $\vee_{p \leq a} a_p \leq a$. Supposing that $q \in Cop(M)$ such that $q \leq a_p$, then $\{a\} \times \lhd(q) \leq \hat{\pi}_\tau(H_p) \wedge \hat{\pi}_M(a_p) \leq T$ for every $p \in Cop(M)$. Thus $p' \leq a$ whenever $q \leq a_p$. That is, $a_p \leq a$ for every $p \leq a$. Hence $\vee_{p \leq a} a_p \leq a$ and the equality $\vee_{p \leq a} a_p = a$ obtained. As $a \in H$ and $H \in \sigma(\tau)$ there exist finite co-prime elements $p_1, \dots, p_n$ with $p_i \leq a$ for any $i = 1, \dots, n$ such that $a_{p_1} \vee \dots \vee a_{p_n} \in H$. Let $\tilde{H} = H_{p_1} \wedge \dots \wedge H_{p_n}$. It remains to show that $\tilde{H}$ has the necessary conditions.

It is clear that $\tilde{H} \in \varphi$ and $a \in \tilde{H}$. Now we prove that $\tilde{H} \subseteq H$. For this goal we show for every $b \in \tau$, if $b \in \tilde{H}$, then $b \in H$. Let $b \in \tilde{H}$. Then for every $q \in Cop(M)$ that

$q \leq a_{p_i}$, we have $\{b\} \times \lhd(q) \leq \hat{\pi}_\tau(H_{p_i}) \wedge \hat{\pi}_M(a_{p_i}) \leq T$ for $i = 1, \ldots, n$. Therefore $q \leq b$ whenever $q \leq a_{p_i}$. That is, $a_{p_1} \vee \ldots \vee a_{p_n} \leq b$. Since $a_{p_1} \vee \ldots \vee a_{p_n} \in H$ and H is an upper subset of τ, it follows that $b \in H$. So $\tilde{H} \subseteq H$. Hence $H \in \varphi$ and $\sigma(\tau) \subseteq \varphi$.

Now our task is to show for another topological space $(\rho(\tau), \tilde{\varphi})$ and any admissible topology φ over τ that $\tilde{\varphi} \subseteq \varphi$, $\tilde{\varphi} \subseteq \sigma(\tau)$. Let $\vee a_i \in H$ and $\vee a_i = \tilde{a}$. Set $A_i = \{a \in \tau | a_i \leq a\}$ for any i and $K = \left\{A \subseteq \tau | \exists i_1, \ldots, i_n, \cap_{j=1}^{n} A_{ij} \subseteq A\right\}$. So K is a filter generated by $\{A_i\}$ over τ. Now we make φ' as another topology on τ. Let $p \in \varphi'$ for $p \subseteq \tau$ if and only if either $\tilde{a} \notin p$ or $p \in K$. In the following, we show that T is open in $(\rho(\tau), \varphi') \otimes (M, \tau)$. Let $\{a\} \times \lhd(p) \leq T$ for every $l \in \tau$ and $p \in Cop(M)$. Then $p \leq a$. We have two cases: $a = \tilde{a}$ or not. First, let $a \neq \tilde{a}$ and set $B = \{b \in \tau | a \leq b\} - \{\tilde{a}\}$. It is obvious that $B \in \varphi'$ and $\{a\} \times \lhd(p) \leq \hat{\pi}_\tau(B) \wedge \hat{\pi}_M(a)$. On the other hand, if we suppose $\{b\} \times \lhd(q) \leq \hat{\pi}_\tau(B) \wedge \hat{\pi}_M(a)$ for any $b \in \tau$ and $q \in Cop(M)$ then $\{b\} \times \lhd(q) \leq T$. Therefore $\hat{\pi}_\tau(B) \wedge \hat{\pi}_M(a) \leq T$ and also T is open in $(\rho(\tau), \varphi') \otimes (M, \tau)$.

Now let $a = \tilde{a} = \vee a_i$. Then there exists an i such that $p \leq a_i$ and since $a = \vee a_i \in A_i$ for each i, we have $\vee a_i \in \cap A_i$ and also $\cap A_i \in K$. Hence $\cap A_i \in \varphi'$ and $\{a\} \times \lhd(p) \leq \hat{\pi}_\tau(\cap A_i) \wedge \hat{\pi}_M(a_i)$. Now let $\{b\} \times \lhd(q) \leq \hat{\pi}_\tau(\cap A_i) \wedge \hat{\pi}_M(a_i)$ for arbitrary $b \in \tau$ and $q \in Cop(M)$. Then $\{b\} \times \lhd(q) \leq T$. Therefore $\hat{\pi}_\tau(\cap A_i) \wedge \hat{\pi}_M(a_i) \leq T$.

The topology φ' is admissible and $\tilde{\varphi} \subseteq \varphi'$, because T is open in $(\rho(\tau), \varphi') \otimes (M, \tau)$ and the set $\{\{a\} \times \lhd(p) | a \in \tau, p \in Cop(M)\}$ is a join generating base for $\rho(\tau) \otimes L$.

Since $H \in \tilde{\varphi}$ and $\tilde{a} \in H$, then $H \in \varphi'$ and $H \in K$. So, there exist finite elements $i_1, \ldots, i_n$ such that $A_{i_1} \cap \ldots \cap A_{i_n} \subseteq H$. That is, $a \in H$ when $\vee_{i=1}^{i=n} a_i \leq a$ for every $a \in \tau$. Hence $H \in \sigma(\tau)$ and $\tilde{\varphi} \subseteq \sigma(\tau)$.

Recall that **Top** is a full reflective and co-reflective subcategory of **TopFuzz** *via* full and faithful power functor ρ : **Top** $\to$ **TopFuzz**. According to the previous Lemma, the mapping $\oplus id_\tau : \oplus\{(\tau, \varphi) | \varphi \text{ is admissible topology over } \tau\} \to (\tau, \sigma(\tau))$ is a quotient map and since the functor ρ : **Top** $\to$ **TopFuzz** preserves co-limits, it follows that the mapping $e = \rho(\oplus id_\tau) : \prod\{(\rho(\tau, \varphi) | \varphi \text{ is admissible topology over } \tau\} \to (\rho(\tau), \sigma(\tau))$ is a regular epimorphism in **TopFuzz**.

Lemma 1.22 A fuzz space (F', τ) is core compact, if $() \otimes (F, \tau)$ preserves the co-equalizers in **TopFuzz**.

Proof: $() \otimes (F, \tau)$ preserves regular epimorphism in **TopFuzz**, because it preserves the co equalizer in **TopFuzz**. So the mapping $e \otimes i_d : \prod(\rho(\tau), \varphi_\alpha) \otimes (F, \tau) \to (\rho(\tau), \sigma(\tau)) \otimes (F, \tau)$ is a regular epimorphism too. Since φ_α is admissible for any α, T is open in $(\rho(\tau), \varphi_\alpha) \otimes (F, \tau)$. But $(\prod(\rho(\tau), \varphi_\alpha)) \otimes (F, \tau) \cong \prod((\rho(\tau), \varphi_\alpha) \otimes (F, \tau))$. So $(e \hat{\otimes} i_d)(T)$ is open in $(\prod(\rho(\tau), \varphi_\alpha)) \otimes (F, \tau)$. By Corollary 2, T is open in $(\rho(\tau), \sigma(\tau)) \otimes (F, \tau)$ and also from Lemma 2, (F, τ) is a core compact fuzz space.

Let $S = \{0, a, 1\}$ such that $0 \leq a \leq 1$. If $0' = 1$, $a' = a$, $1' = 0$, and $\tau = S$, then $(S, ', \tau)$ is a fuzz space.

Theorem 1.23 Let $(M, ', \tau)$ be a fuzz space. Then the following are equivalent:

1. (M', τ) is exponentiable in **TopFuzz**.

2. $(M, ', \tau)$ is core compact.

3. The mapping $ev : \left(F_1^{M'}, \eta\right) \otimes (M', \tau) \to (F_1', \tau_1)$ defined by $ev(D) = \vee\{f(x) | f \in A, (A, x) \in D\}$ is continuous for every fuzz space (F_1', τ_1).

4. For $(F_1, ', \tau_1) = (S, ', \tau_1)$, where $\tau_1 = S$, part (3) holds.

Proof: According to "special adjoint functor theorem" in [13], (M', τ) is exponentiable in **TopFuzz** if and only if $() \otimes (M, \tau)$ preserves co-products and co-equalizers. Since $\prod L_i \otimes L \cong \prod (L_i \otimes L)$ is always true for fuzz spaces, to prove (1) $\Rightarrow$ (2) it is enough to show if $() \otimes (M, \tau)$ preserves co-equalizers in **TopFuzz**, then (M', τ) is core compact and it is proved in Lemma 2. The implication (2) $\Rightarrow$ (3) is just Lemma 2. (3) $\Rightarrow$ (4) is clear.

Now in this part we prove (4) $\Rightarrow$ (2). First we introduce $S^M : [M \to S] \cong \tau$ by $f \mapsto \hat{f}(a), f \prec g$ if and only if $f \leq g$ if and only if $\hat{g}(a) \leq \hat{f}(a)$. Therefore $Low_{\precsim}[M \to S] \cong Low(\tau)$, where $Low(\tau)$ is the lower subsets of τ. The Isbell topology over S^M is generated by $T(H, a)$ for $H \in \sigma(\tau)$, where $T(H, a) = \cap\{I \in S^L | I \cap H \neq \varnothing\}$. So the Isbell topology over S^M is $\{T(H, a) | H \in \sigma(\tau)\}$. The evaluation mapping $ev : S^M \otimes M \to S$ defined on every generator $\downarrow x \times \lhd (p)$ for $x \in \tau$ as follows:

$$ev(\downarrow x \times \lhd (p)) = \begin{cases} a & p \leq x \\ 1 & p \nleq x \end{cases} \tag{4}$$

Now let $p \in Cop(M)$ and $x \in \tau$ such that $p \leq x$. Then $ev(\downarrow x \times \lhd (p)) = a$ and $\downarrow x \times \lhd (p) \leq \hat{ev}(a)$. Since $\hat{ev}(a)$ is open, then there exist $H \in \sigma(\tau)$ and $y \in \tau$ such that $\downarrow x \times \lhd (p) \leq \hat{\pi}_{S^M}(T(H, a)) \wedge \hat{\pi}_M(y) \leq \hat{ev}(a)$. So $x \in H$ and $p \leq y$. To prove $y \leq \wedge H$, it suffices to check that $y \leq x_1$ for every $x_1 \in H$. Let $x_1 \in H$ and p_1 be an arbitrary element in $Cop(M)$ and $p_1 \leq y$. Then $\downarrow x_1 \times \lhd (p_1) \leq \hat{\pi}_{S^M}(T(H, a)) \wedge \hat{\pi}_M(y)$. Therefore $\downarrow x_1 \times \lhd (p_1) \leq \hat{ev}(a)$ and $p_1 \leq x_1$. In fact we have shown $y \leq x_1$ for every $x_1 \in H$, that is, $y \leq \wedge H$. According to Lemma 2, (M', τ) is a core compact space in **TopFuzz**.

3. Conclusions

It is well known that the category **TopFuzz** of all fuzz spaces is both complete and co-complete and some categorical structures of it have been introduced by many authors. Since the category **Top** of all topological spaces is a reflective and co-reflective subcategory of **TopFuzz**, it follows that **TopFuzz** is not a Cartesian closed category. In this note, we have introduced some characterizations of the exponentiable fuzz spaces. One important result we have proved is that a fuzz space (F', τ) is exponentiable if and only if it is core compact, that is $a = \vee\{b \in \tau | b \ll a\}$, for every $a \in \tau$.

Author details

Ghasem Mirhosseinkhani
Department of Mathematics and Computer Sciences, Sirjan University of Technology, Sirjan, Iran

*Address all correspondence to: gh.mirhosseini@yahoo.com; gh.mirhosseini@sirjantech.ac.ir

References

[1] Wang GJ. Theory of topological molecular lattices. Fuzzy Sets and Systems. 1992;**47**:351-376

[2] Wang GJ. Order-homomorphisms on fuzzes. Fuzzy Sets and Systems. 1984;**12**: 281-288

[3] Wang GJ. Generalized topological molecular lattices. Scientia Sinica. 1984; **8**:785-798

[4] Akbarpour M, Mirhosseinkhani GH. Exponentiable objects in some categories of topological molecular lattices. Hadronic Journal. 2017;**40**:327-344

[5] Akbarpour M, Mirhosseinkhani GH. Cartesian closed subcategories of topological fuzzes. Algebraic Structures and Their Applications. 2019;**61**:23-33

[6] Demirci M. Category theoretic fuzzy topological spaces and their dualities. Fuzzy Sets and Systems. 2013;**227**:1-24

[7] Hutton B. Products of fuzzy topological spaces. Topology and its Applications. 1980;**11**:59-67

[8] Hutton B, Reilly I. Separation axioms in fuzzy topological spaces. Fuzzy Sets and Systems. 1980;**3**:93-104

[9] Li YM, Li ZH. Top is a reflective and co-reflective subcategory of fuzzy topological spaces. Fuzzy Sets and Systems. 2000;**116**:429-432

[10] Yang ZQ. The cartesian closedness of the category fuzz and function spaces on topological fuzzes. Fuzzy Sets and Systems. 1994;**61**:341-351

[11] Li YM. Generalized (S,I)-complete free completely distributive lattices generated by posets. Semigroup Forum. 1998;**57**:240-248

[12] Fan TH. Product operations in the category of molecular lattices. Chinese Science Bulletin. 1987;**32**(8):505-510

[13] Adamek J, Herrlich H, Strecker GE. Abstract and Concrete Categories. New York: John Wiely and Sons Inc.; 1990

[14] Escardo M, Lawson J, Simpson A. Comparing cartesian closed categories of (core)compactly generated spaces. Topology and its Applications. 2004;**143**: 105-145

[15] Escardo M, Heckmann R. Topologies on spaces of continuous functions. Topology Proceedings. 2001-2002;**26**(2): 545-564

[16] Li YM. Exponentiable objects in the category of topological molecular lattices. Fuzzy Sets and Systems. 1999; **104**:407-414